# Through Eugène Dubois' eyes

# Through Eugène Dubois' eyes

## Stills of a turbulent life

*By*

Paul C.H. Albers
John de Vos

TUTA SUB AEGIDE PALLAS · 1683 ·

BRILL

LEIDEN • BOSTON
2010

This book is printed on acid-free paper.

ISBN 978 90 04 18300 1



PRINTED IN THE NETHERLANDS

# Contents

## Preface

I can hardly describe my delight and surprise when I first opened a series of emails from my long-time friend and colleague, John de Vos, containing the manuscript of "Through Eugène Dubois' Eyes". When I was researching and writing my biography of Dubois, John's knowledge of and insight into Dubois' fossil collections, archives, and photographs was of immeasurable help. In fact, it was John who first interested me in writing a biography of Dubois by apparently casually making me aware of the rich archives at Naturalis and asking if I cared to "do something" with them. That simple query lead to my many years of obsession with Dubos that culminated in "The Man Who Found the Missing Link". Dubois has been, to me, a tremendously fascinating man whose life story illustrates both the best and perhaps the worst of the results of a scientific obsession. In turn, Dubois and his views invaded my mind in a way I found both disturbing and rewarding.

I became immersed in Dubois' life and thinking. Three examples will show what I mean. Once I was challenged by an alert copyeditor, who remarks that I was writing sentences in English that sounded as if they had first been written in Dutch. Another time, on holiday, I picked up a rather nondescript brownish-grey hat and put it on, asking my husband if he liked it. "That's Dubois' hat," he remarked. "The one in his last photograph." Indeed, without realizing what I was doing, I had selected a hat that very closely resembled Dubois'. On the third occasion, when I was due to give a lecture about Dubois, Ian Tattersall of the American Museum of Natural History introduced me. He commented that my approach to biography was somewhat unconventional – I wrote Dubois' biography as if it were a novel, because that seemed to me to be the best way to convey his life and character. Tattersall said that, of course, we could never know what had transpired at particular occasions in Dubois' life nor could we know what he has said or thought. My first response – a response that I barely prevented myself from giving out loud – was : "Don't be silly! Of course we know. I was *there*!" Only "I", in this case, was Dubois, not me. I am told this kind of intense identification often happens to biographers. I can only confess it happened to me, with Dubois.

To receive this manuscript from John de Vos and Paul Albers was a completely unexpected bonus. I realize now that, when I examined the glass plate negatives of Dubois' life, some were unavailable as they were being conserved. I had forgotten this fact and so was especially thrilled to see pictures of many places, fossils, and people that I had not encountered in photographs of Dubois' life. Most particularly, I was delighted to "meet" Adam Prentice, Dubois' close friend in the Dutch East Indies (now Indonesia) who, though no scientist, providing Dubois with a sounding board as he developed his ideas about the extraordinary fossil he found. Prentice was in many ways the key to my understanding of Dubois.

It must be remembered that Dubois accomplished the near-impossible in his fossil-hunting. He worked out from first principles where the ape-man or "missing link" ought to have lived, and where it might have been preserved as a fossil – and went and found it. He

had little help from geological surveys, which had barely been undertaken, or from aerial photographs, satellite imaging, radiometric dating, or any of the other technological tools fossil-hunters use today. He had no airplanes, no rugged four-wheel-drive vehicles, and no fat government grants although he was singularly successful in persuading various Dutch colonial officials to grant him time and labor for excavations. He had no medicines to keep himself, his crew, or his family healthy at a time when the deaths from tropical diseases were appallingly high. In fact, one of his own children died at birth in the Indies, which can only have been a cruel sacrifice for him and his family.

In addition to finding his "missing link", Dubois invented new techniques for photographing and measuring his fossils and for comparing them to other species. He had no library full of similar treatises, no comparative anatomical collections aside from those items he could beg, borrow, or buy himself. He set the standard and was sometimes criticized for what he did not do, but it would be more fair to remember that he worked in complete isolation from learned colleagues and the resources available in Europe. To do what he did was astonishing; to expect him to have done more is perhaps unrealistic.

Aside from filling in important gaps in the photographic record of Dubois' life and travels, the images reproduced here capture stunning portraits of life in the colonial world. The people, the landscapes, the villages, the geological exposures, and the vegetation are preserved for historians and others in this volume. It is one thing to read collections of colonial letters home and to imagine the life European colonists led; it is another entirely to see for oneself those lives, the clothing, the houses, the roads, the servants, and the elephants.

Finally, the newly-found correspondence between Dubois and the various editors of Brill, the intended publisher of his treatment of the Trinil fauna, is almost painful to read. It is clear that the Dubois Collectie, as it is now known, is of tremendous importance to science and its full potential has not yet been exploited. But the correspondence shows most clearly Dubois' perfectionist nature and his inability or unwillingness to hand over his treasures to the scientific world in writing. He could not finish the work; indeed he hardly began it, and frequently turned his active mind and broad abilities to other projects that were not quite so close to his heart.

If I must summarize what this book offers, it is *more*: more images, more letters, more details, more insight into the workings of a brilliant but unquestionably difficult man of science. We shall not see Dubois' like again so it is doubly fortunate that Albers and de Vos have uncovered so much more about his life.

Pat Shipman
State College, PA.
USA
October 27, 2009

**Note from the authors**

This book is not intended to be a complete biography on the life of Eugène Dubois. Bert Theunissen and Pat Shipman have preceded us on that path and we will not present more than a short recapitulation of their previous work. We more or less consider, you, the reader of this book, to be acquainted by the knowledge that Eugène Dubois became famous for being the first man to find *Pithecanthropus erectus* (nowadays *Homo erectus*), at that moment in time a missing link between man and ape. The circumstances of his life where extraordinairy enough to have already filled a Ph.D thesis and a biography by the authors mentioned above, but the sheer amount of interesting data still leaves room to add more to his tale. We will just try to fill in some of the gaps we stumbled upon, and will do so by presenting you the outlays of a book that Dubois started very early in his career, in his mind surely already in Indonesia, but which he never finished. Though this fact itself was known, the real impact of this never finished work became only apperent to us when we discovered some 40 hithertho unknown letters directed to Dubois in the archives of his intended publisher, Brill in Leiden.

But next to that we also present an ample choice out of the huge photographic collection in the Dubois archives at Naturalis, the National Museum of Natural History of the Netherlands, all material that Dubois either (had) made or acquired. Out of that collection we particularly chose those photographs that have a story to tell that you might not have heard before or that illustrates what we consider of special (scientific) interest.

Because in the end, that is our ultimate goal: to disclose for you these pictures, which need not be introduced as they will speak for themselves…

Paul Albers & John de Vos

"Le mieux c'est l'ennemi du bien"

words that Frans de Stoppelaar uttered regularly to persuade Dubois to finish his book.
(The better is the enemy of the good)

# Introduction

Although the first finds of fossil hominids date back to 1891, thinking about evolution of Man started at least as early as 1844 when Robert Chambers anonymously published his book '*Vestiges of Natural History of Creation*', in which he presented a development theory. Chambers did not stress the point, but his development hypothesis clearly made Man an immediate descendant of the apes. The anatomist Richard Owen used his expertise to disprove the theory of evolution at its most controversial point –man's link with the apes by pointing at the heavy eye-brows of the great apes, which were missing in modern Man. As the eyebrows are independently developed, nor influenced by inner or outer factors, Man must have, if Man was descendent from the great apes, heavy eyebrows; and that, he pointed out is not the case.

However, a decade later in the Neanderthal near Düsseldorf a skull was found with heavy eyebrows. The fossil came into the hands of Hermann Schaaffhausen, professor of anatomy at the University of Bonn, who was convinced that the remains were very old and hominid. Their strange morphology was caused by deformation, but the oligocephalic form of the skull was, according to Schaaffhausen, not comparable to any modern race, not even with the most 'barbarian' races. The heavy eyebrows, characteristic for great apes, were according to Schaaffhausen typical for the Neanderthal. The skull therefore must have belonged to an 'original wild race of North-western Europe'. Some even considered it as the skull of an idiot, an 'old Dutchman' or a Cossack.

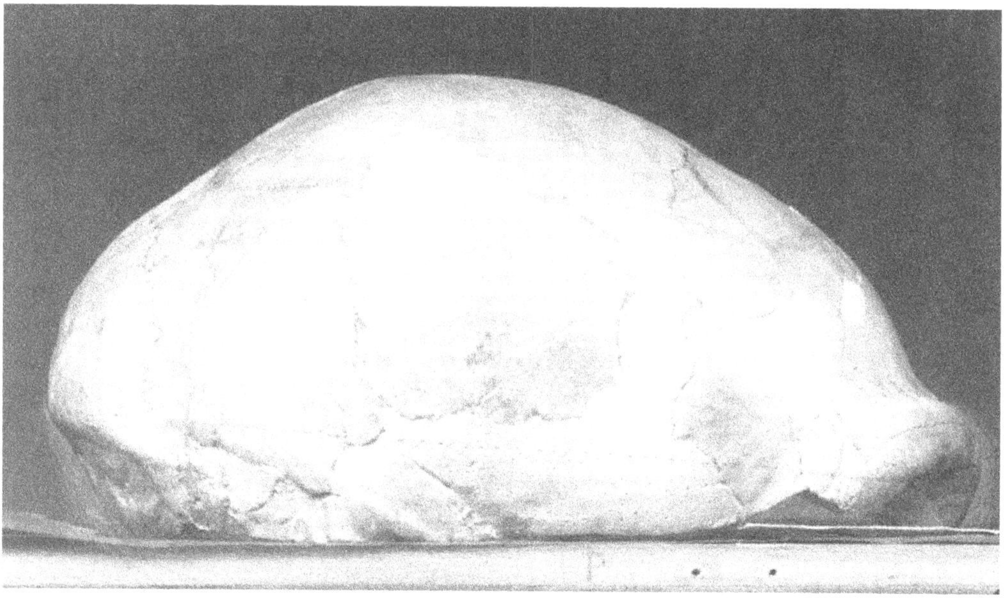

DUBO1426 Skull of Spy (*Homo neanderthalensis*) uncovered in 1887

DUBO4103 Thomas Huxley

Charles Darwin published his famous book '*On the origin of species*' in 1859, and in it he set out a theory of evolution characterised by a gradual development in which natural selection is the thriving force. Although Darwin at that point chose not to mention the subject but ever so briefly, others were soon drawing the conclusion that there is no separation between Man and apes. In this context the Neanderthal skull became central in the discussion. Two ideas came up, which are still debated today: (1) Neanderthal Man belongs to recent Man; (2) Neanderthal Man is a species of its own. It was Thomas Huxley who set the tone of the discussion by describing the morphological characteristics of the Neanderthal as primitive, yet definitely human. He also pointed to the large brain capacity as proof of the Neanderthal's human nature. William King, professor of anatomy at Queens College (Ireland), in contrast considered, without giving scientific arguments, Neanderthal Man as a new species and called it *Homo neanderthalensis*. Later finds were also generally ascribed to a (primitive) race of modern humans, such as the mandible of La Naulette and the Spy skeletons.

Slide that Dubois used in class of the hypothesized *Pithecanthropus alalus* drawn by Gabriel Max

For Ernst Haeckel fossils were no longer required to prove that Apes were part of the evolution of Man, because the process could be proven already by anatomy and embryology. He introduced the name *Pithecanthropus* to identify the missing link between Ape and Man. The species he called *Pithecanthropus alalus,* the ape-man without speech.

## *Dubois and his Pithecanthropus*

Marie Eugène François Thomas Dubois was born in 1858, a year before Darwin published his *Origin of species*, and two years after the discovery of the first Neanderthal skull. He grew up in a period that witnessed the rapid acceptance and dissemination of the theory of evolution. During his youth the problem of human ancestry was central to many discussions on evolutionary theory. And by the end of his university studies opponents and supporters of an evolutionary ancestry for humans still agreed that no hominid fossils were known to provide proof of human evolution. Dubois was to be among the first to bring about a change in this climate of opinion. Born in Eijsden in the south of the Netherlands, near St. Peter's Mountain where in cretaceous rocks the remains of mosasaurs had been found since 1766, Dubois had been interested in palaeontology from his early childhood. Anecdotal stories of him roaming around in the area, examining deep pits are still being rcpeated. He was born in a Roman Catholic family and had two sisters: Marie, later sister Marie-Angelique, and Gerardine and he also had one brother, Victor. His father was a pharmacist and Mayor of Eijsden. From his father he learned the Latin names of the plants and his mind was set for a scientific career. He was sent to a 'far' away public school as opposed to the more usual and nearby Catholic institutions, and was more likely to have heard there from the discussions and topics in evolutionary theory.

Dubois' house of birth in Eijsden

Heated discussions surrounding the ideas of the German anthropologist Carl Vogt, who in the late 1860s lectured in the Netherlands on evolutionary theory and human descent, probably furthered his interest in paleo-anthropology. From 1877 to 1884 Dubois studied medicine at the University of Amsterdam; again an institute were Catholic upper class boys were not likely to go. But he did well and subsequently became a Reader in anatomy in 1886. Although he now was on the brink of a prosperous academic career, his predilection for paleo-anthropology made him decide to change course: more and more he became possessed by the idea of beginning a search for hominid fossils that might prove human evolution.

The Dubois family

But where should he start looking? Dubois referred to Darwin, Wallace and Lydekker to explain this choice. In his *Descent of Man* (1871) Charles Darwin had reasoned that our human ancestors must have lived in the tropics, since human beings had lost their fur pelts in the course of their development. He had suggested Africa, where chimpanzee and gorilla live, as the most probable region of human origins. Alfred Wallace, on the other hand, had stressed the importance of searching for the ancestors of present-day anthropoids in caves and tertiary deposits in both Africa and Southeast Asia. Finally Richard Lydekker had described in 1879 a primate fossil - an incomplete jaw with a number of teeth - from the Siwalik Hills in British India (now Pakistan), which seemed to throw some light on human descent. According to Lydekker, this primate, whom he named *Palaeopithecus sivalensis*, could be regarded as a predecessor of the chimpanzee (*Anthropopithecus*). Yet, he added that the fossil also showed resemblance to both gibbon and human. Dubois concluded from this that "the Gibbon group which in earlier geological periods had developed further" might have played a role in human evolution.

For Dubois the East Indies seemed a suitable area, the more so because this colony of the Netherlands lay wide open to him. And also Karl Martin, geologist at Leiden University had published on the presence of a Siwalik fauna in Indonesia. Adding all this up, the idea to start looking in Indonesia was not as farfetched as it seems now. Following this call, he gave up his position at the university and in order to sustain himself he joined the Dutch East Indies Army as a medical officer, and boarded the SS Amalia with his young wife Anna Lojenga and their firstborn daughter Eugenie to begin his search for the missing link in the Dutch East Indies.

In December 1887 he arrived in Padang on Sumatra. His duties as medical officer initially took much more of his time than he had hoped and it took him quite some time before he could start his geological investigations in earnest. In May 1888, he was seconded to Pajakombo in the Padang Highlands, where he began a search for caves suitable for paleontological excavation. By August he found some proof that fossil mammals were actually to be found in the caves of Sumatra. Meanwhile Dubois had also set up a lobby of people that supported his excavation plans amongst which Karl Martin was the most influential voice. Eventually, with the support of this lobby, the colonial government relieved Dubois from his army-duties and enabled him to dedicate all his time to this search and actually put him in charge of carrying out paleontological excavations on Sumatra and, if necessary, on Java.

DUBO2115 Dubois was sent this picture by a fellow physician in Indonesia to acquaint him with elephantiasis

The Sumatran cave fauna, however, soon proved too young to include any human forerunners. Therefore, in 1890, Dubois decided to continue his excavations on Java, where fossils of supposedly Tertiary age had already been found by, among others, Franz Junghuhn and Radèn Saléh. Moreover, in 1888, the mining engineer Van Rietschoten had found a fossil skull near the village of Wajak on Java, which though fully human, clearly differed from the modern Javanese population. Java thus held promising prospects.

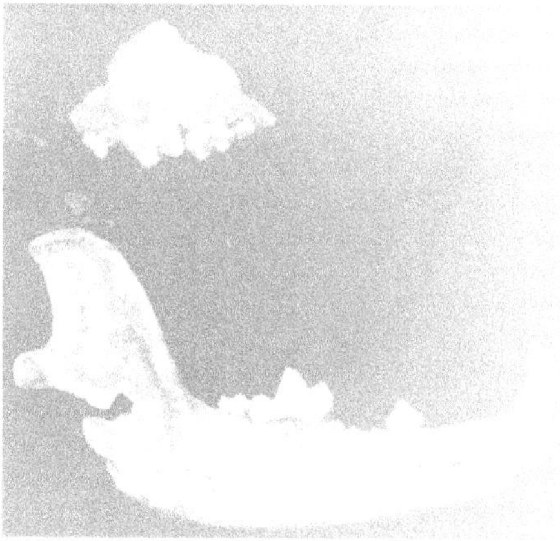

DUBO1092. Mandible and maxilla of the dog from Hoekgrot.

At the Hoekgrot ('Corner cave'), very near to where the Wajak Skull was found, he found a red painted human skeleton. In addition to this skeleton he found a partial skeleton of a dog. Java thereby became his main settlement, but it is a very large island indeed, and Dubois did not restrict himself to one site, not even after he had found *Pithecanthropus.* Other material was notably collected in Kedung Brubus, Butak in the Kendeng Hills, the Patiayam, in the north of Java, and, most importantly in Trinil, within a bent of the Solo-river.

When we write that Dubois found something somewhere, we should make clear that probably 95% or more was not actually found by him at all, nor in his presence. The actual work was done by two members of the engineering corps that were assigned to him, the sergeants Gerard Kriele and Anthony de Winter, along with 25-50 forced labourers (convicts) to help him with the excavation work. They were enduring all the hardship whilst Dubois was mostly at his home studying the fossils that were being transported to him.

The letters of Kriele and De Winter reveal that it was no fun and games and their circumstances were downright miserable. Local governors were not always pleased to see them start working in their neighbourhood as they caused them extra work. Villagers were instructed to misinform them and throw away fossils if they found them. Forced labour was the only possibility to get the work done, payed labour ran off after the first day not wanting to do the hard work required for the income offered. The convicts, whose conditions were obviously even more miserable, were regularly insubordinate, lazy, run off, bribed the overseers that came with them, stole fossils or deliberately destroyed them, and were often ill and even died. Kriele and de Winter often handled the pickaxes themselves as they did not trust their workers to be careful enough when digging in layers where bones were to be expected. And they had their share of illnesses too: frequent fever,

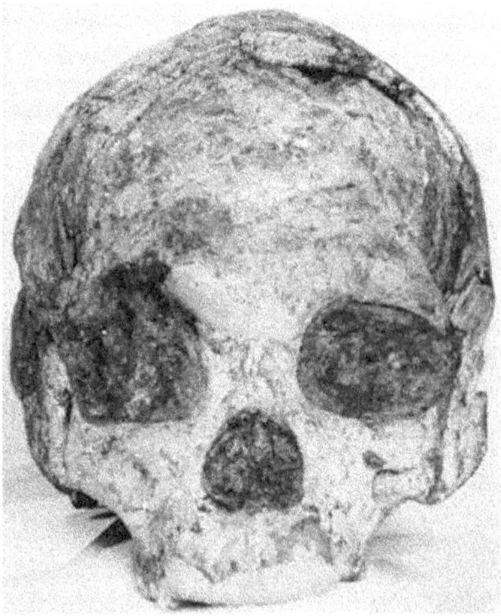
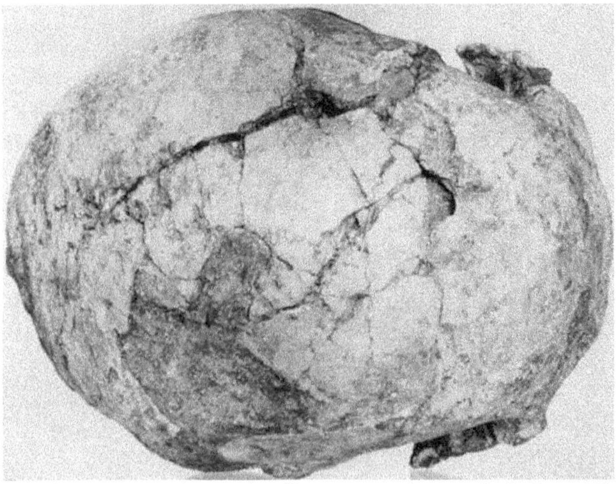
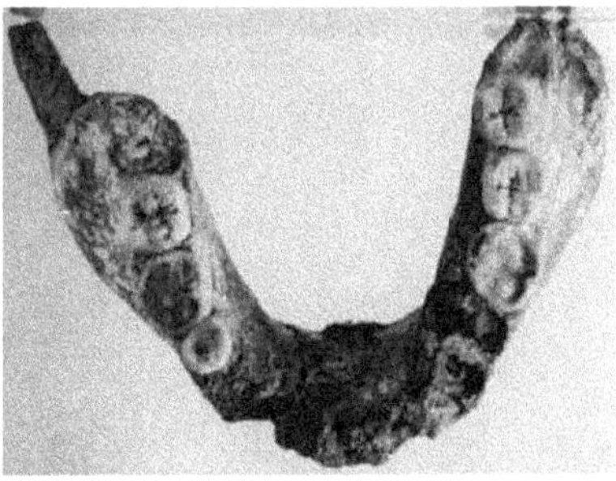

DUBO4302/3/4/7 Skull found near the village of Wajak on Java, here on photographs that were only recently discovered and which must have been taken before extensive 'restauration' on this skull had taken place

diarrhea, sunstroke and callous growths on their feet. There is also mention of 'women's disease' which no doubt translates to some sexually transmitted disease. They were probably in luck that Dubois was a medic and could supply them with medications, as many a letter ends with such a request.

They were also constantly begging him for money, being ripped off by the locals at every occasion and having to pay for their guides and overseers, materials, equipment and rent for their housing and the convicts'; they were regularly complaining they had not enough money left to buy food. Some of that might have had to do with their drinking habits, or at least de Winters' drinking habits, but the circumstances being as they were, one should say in their favour, that there are less worse circumstances that can make someone start drinking. Their dedication to the job and to Dubois was however enormous and Dubois returned their affection to him with a continued support e.g. writing letters of recommendation, even after he had left Indonesia for good.

DUBO0906 Anthony de Winter

DUBO0911 Gerardus Kriele

(On these photographs they are posing in new uniforms that had just been issued in 1894.)

After cave exploration in Java proved unrewarding Dubois now turned his attention and workforce to the open field. Success was almost immediate.

In November 1890, near Kedung Lumbu at Kedung Brubus in the Kendeng Hills, he found a fragment of a mandible that he described as follows: "Amidst the remains of typical representatives of the fauna concerned, and in the same layer ..... a human fossil was found, the right side of the chin of a lower jaw with the sockets of the canine tooth and of the first and second premolar....[T]his fossil [jaw] forms a different and probably lower type than any previously known".

DUBO3985 *Pithecanthropus erectus* jaw fragment from Kedung Brubus found in 1890

In August 1891, during the second digging season in Java, Dubois began excavations at Trinil, a locality that was to acquire historical significance. (Most photographs of the site were made in 1900; amongst them the one on the cover of this book.)

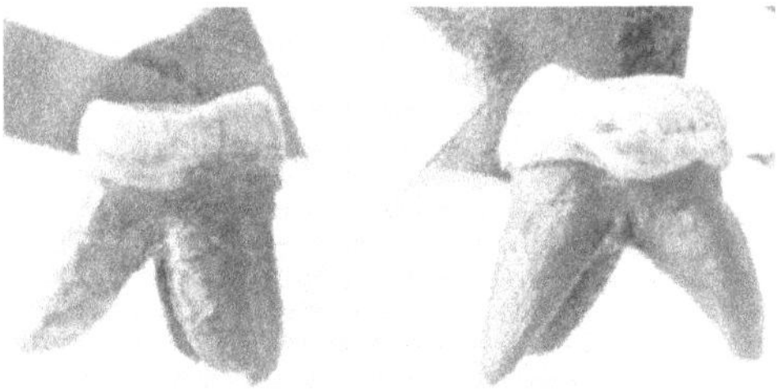

DUBO0318 *Pithecanthropus erectus* molars

An enormous number of vertebrate fossils were unearthed, and in September 1891, the first remains of a primate, a third molar, emerged from the sediments. At first, Dubois referred the fossil to a chimpanzee (*Anthropopithecus*). In one of his reports to his superiors he wrote: “This genus of anthropoids, occurring only in West- and Central-equatorial Africa today, lived in British India in the Pliocene and, as we can see from this discovery, during the later Pleistocene in Java”.

DUBO1303 Skull cap of *Pithecanthropus erectus* photographed before the matrix was removed from the skull cavity

The skull cap for which Dubois acquired fame was unearthed in October. In Dubois' opinion it was clearly distinguishable from that of the orang-utan and the gorilla. It had to belong to the same fossil chimpanzee from which a molar had been found a month earlier. Despite having designated the find as a ‘chimpanzee’, Dubois was well aware that he had made an important discovery. The Javanese cranium was higher and larger than that of the recent chimpanzee and substantially more human-like than any known anthropoid, whether fossil or recent. He does however not have a chimpanzee skull to compare with and writes to his friend and colleague Max Weber in the Netherlands to send him one. Weber tries his utmost but chimp skulls are not easy to get a hold of and Dubois has to postpone his conclusions.

Meanwhile in August of the following year a third primate fossil was discovered, this times an almost completely preserved left femur. "This thigh bone," Dubois stated, "lay at the same level in which both the other parts were found, yet following the direction of the earlier stream which deposited the tuff material 15 m upstream. From the circumstances of the find and [my] comparative research it is evident that the three skeletal elements belong to one and the same individual, probably a very aged female". The almost perfectly human characteristics of the femur indicated that the individual must have walked upright, and this induced Dubois to christen his "chimpanzee" *Anthropopithecus erectus.* Further he stated: "In view of all three skeletal elements, especially the femur, *Anthropopithecus erectus* EUG. DUBOIS approaches modern Man more closely than any of the three great apes, a fact which is in harmony with the thesis of Lamarck, and later of Darwin and others, that the first step in the direction of humanisation of our ancestors was the [acquisition of the] erect position".

In December 1892 the chimpanzee skull that he so desperately needed for comparison finally arrives and soon additional investigation of the remains convince Dubois that they represent an intermediary form between humans and apes. He therefore decides that it is more appropriate to designate his find an "ape-man", *Pithecanthropus*, instead of a "man-ape", *Anthropopithecus* (the name *Pithecanthropus* was coined by Ernst Haeckel for the then still hypothetical link between humans and fossil apes). In 1894 Dubois published the results of his studies under the title "*Pithecanthropus erectus*, eine menschenaehnliche Uebergangsform aus Java".

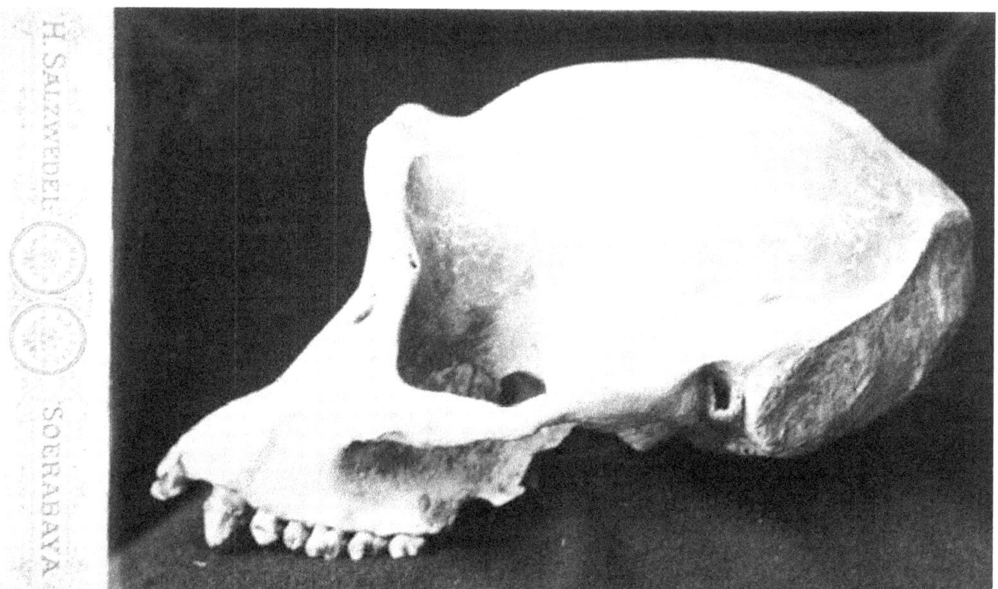


DUBO2113 print of a photograph made whilst in Indonesia of the chimpanzee skull that Dubois acquired to compare his *Pithecanthropus* to.

Dramatic high in the Dubois archives:
the manuscript in which he changes *Anthropopithecus erectus* into *Pithecanthropus erectus*

During the excavations Dubois stayed in Tulung Agung where he had a house. His wages afforded him several servants, a cook and a nanny. There are just very few photographs with people on them and within the Naturalis collection there are none of his family whilst he was in Indonesia (there is at least one in the family collection shown further on in this book). There are a few of his staff, as of his friend Robert Boyd, and one other portrait, most likely of Adam Prentice. The amount (30.000+) of fossils was so large that he stored them on his veranda (see below: DUBO0910).

DUBO0907 Another picture of the 'fosfilled' veranda

DUBO0892 Boy servant at Tulung Agung DUBO0897 Man servant at Tulung Agung.
The tree is a papaya, a species carrying its fruit on the stem (caulifloor).

DUBO0168 Bathing facility(?) in nearby brook.

None of the portraits hold names except for the one of Djongos Nassi (DUBO0896). Of most of the portraits we do not have any prints in paper. There is one print of the boy servant (DUBO0892) stating it to have been taken in Tulung Agung. The only other portrait identified without any doubt is the portrait of Robert Boyd, a planter at a plantation called Mringin nearby Tulung Agung, packed in the same box next to the negative of the portrait of Boyd however was another negative of a portrait of a younger man. The negatives are evidently of the same age and suffered the same damage, flaking of the photographic layer possibly due to heat by a lamp coming to close to the negative. Given this 'nearness' we feel confident in assigning the name Adam Prentice to it, who was Boyd's right hand man at Mringin and both became good

DUBO0896 Djongos Nassi

DUBO0901 Robert Boyd

DUBO0900 Adam Prentice(?)

friends of Dubois and exchanged many visits and letters.

We think that the tree depicted on DUBO1314 was probably also photographed at Mringin. It has turned out to be a key picture in the collection for at first we thought it to be a tree in the Netherlands but when we asked a botanist to identify the species he pointed out the tree left in the background was a banana and the leaves directly at the back of the tree did not belong to the tree but were probably of a coffee shrub. The tree itself, of which just a few leaves are visible in the top right corner of the photograph, is most likely a *Cynometra ramiflora,* a relatively rare caulifloor (flowering on the stem) species. The outgrowths on the stem are the spots were flowers have been and where a few

DUBO1314 *Cynometra ramiflora*(?)

seedpods are visible. (When we showed the picture to Pat Shipman she identified a femur of a bovine on the background, to the right of the stem just before the wall, and there are also an atlas and a sacral, just to illustrate that one can only see what you expect to see, other people always see things you had not seen yourself)

Only after it had now been established that this photograph originated from Indonesia, we saw the resemblance between the wall on the background and the wall on the background behind the girl on DUBO0895. Again we had placed this picture first in the Netherlands, in spite of the Indonesian furniture (which many Dutch colonialists took home with them), but now it seemed much more likely that this picture was also taken at Mringin and portrays Robert Boyd's oldest daughter Anna Grace.

Before Dubois was to return to the Netherlands when his tour of duty ended in 1895, he went to visit the Siwalik Hills, which we will discuss separately further on. After leaving definitively for the Netherlands his sergeants would continue to collect fossils, in particular at the Trinil site until 1900. Dubois never returned to Indonesia, which cannot have been for lack of opportunity. He was to become professor of Geology in Amsterdam and regularly excursions to Indonesia were organized, also from the Geology department. In those days all Dutch universities were very keen on their faraway possibilities and any botanist, zoologist or geologist of any standing would go there and probably even needed to have visits to the colonies on his curriculum vitae to be taken seriously. But Dubois never returned; he had found everything there that he could possibly have hoped to find. No more was to be expected.

DUBO0895 Anna Grace Boyd(?)

*Siwalik*

Before returning to the Netherlands Dubois takes a trip to the Siwalik Hills in the North of what was then British India, now Pakistan. The elaborate trip to the sites where Richard Lydekker had been collecting fossils about twenty years earlier had one main reason. He had to confirm his thesis that there was no 'Siwalik fauna' on Java by comparing the sites and comparing the Indonesian fossils to the ones curated at the Natural history Museum of Calcutta. In particular he wanted to see *Anthropopithecus sivalensis* and is glad he has as he concludes that Lydekker has reconstructed and depicted the jaw incorrectly. We have an account of the trip which was first used by Pat Shipman for her book, but there are also a number of photographs of which she only used one (the elephant). These pictures are intriguing as most of them just depict people, which clearly differ with regard to their social background. These pictures are therefore an interesting document on themselves from an ethnographic point of view. The account of the trip nowhere explicitly mentions that pictures are taken, but descriptions of certain people clearly fit some of the pictures closely. This brings us to another point altogether, the account we have in the Dubois collection itself: there is something odd about it. It is doubtlessly written in different hands, some parts even, as it seems, by a child. It breaths a very distinct suspicion that some form of censorship has been going on, it has been completely rewritten and it ends abruptly, in the middle of a sentence. It might have been constructed from personal letters, or the original might have contained passages otherwise not fit for publication. It seems odd and not Dubois-like to not have kept specific notes on these photographs (as he did do in Indonesia), but the archives do not currently hold them. The account however is an interesting observation on how the British held office in comparison to the Dutch and the differences between Indonesian and Indian cultures. Generalizations are not sparse and people he meets are by no means always favourably judged upon. British drinking habits are a returning subject of wonder as is their pompous upper-class attitude and the ever returning mutton, beef stew and ice cream. In Calcutta he visits an evening party of the Vice-Roy and in officers dress himself gets to see "hundreds of women dressed up for a ball"...

DUBO0799 Elephant owned by the Maharadja of Sirmoor

He is very happy to finally leave Calcutta behind him and longs to see forests, mountains and streams. The crew he hires consists of about ten people, at some points he adds one or two, and as he mentions only men and some photo's of what seems to be (part of) his crew also depict women, these women might not even have been counted. He also hires camels, which he describes as ugly wretched beasts. He rides one of them, for want of try, but climbs down after ten minutes and subsequently decides to walk for an hour "to set his spine straight".

Victor Bruce, 9th Earl Of Elgin
Vice-Roy of India 1894-1899

DUBO0833 Camels in Siwalik landscape

Dubois' crew, depicted below in DUBO0843, can partly be identified by the descriptions we have and the objects they are carrying. Second to right is Dalea, a small man from Madras, and Dubois' main man. Left of him is Allay Deya, the water carrier, identifiable by the goat skin filled with water he is carrying. Almost the furthest to the back, fifth from the right, carrying a broom, is the sweeper, Choonee, who also had to empty the pots, a job for the lowest cast (Dubois describes how selecting the crew takes ages as all these different cast aspects need to be considered). The man most to the left is very possibly a Sikh, judging by the tightness of bindings of his turban. As we know Ram Phal had an army history and the man on the photo to the right is the only one fitting that description, in particular because of the belt he is wearing, we ascribe him that name. Not having been present for the 'group picture' might have been the reason for his solo-portrait.

DUBO0800 Ram Phal was hired to take care of the horses (as he had done for the British military)

DUBO0843

DUBO0798 Probably two Hindu couples with children to the left, the adult male to the right is a Muslem

DUBO0838 Equals DUBO0798 but with two more males to the left, who look much more self-assured than the rest of the group. They are possibly assigned to Dubois' party by the Maharadja of Sirmoor…

DUBO0832 Typical Siwalik landscape

DUBO1276 Siwalik landscape

DUBO1277 Siwalik landscape

DUBO0840 Women

DUBO0801 Child with shoes

Dubois describes in his travel log some aspects of the native clothing. In general the colours are red and white, he says, whereas in Indonesia the common colours are blue and black. He seems amused by the trousers the women are wearing, tight at the ankles, wide at the hips.

According to Dubois the Hindu women are wearing rings as large as bracelets in their noses, and in general many rings, on fingers and toes. When we showed these pictures to experts of this region they were not at all sure these were Hindu women. In those days these groups lived side by side and are not always that easy to identify from pictures. Dubois seems unaware of such subtleties and should not be taken on his word in this matter. To him as to most colonial whites at that time, the natives were principally all 'black'. Dubois was however interested in racial physiology, which may well have been the reason to photograph these people at all. The following pictures, though seemingly of the same type of women, show something interesting on the background. One can just make out an elephant depicted in the rock. The elephant is facing left and its eye is about at the same height as the eyes of the leftmost adult female. The area is renowned for its pictures on rocks but elephants are not a theme, ibexes are.

DUBO0835 Women group in front of elephant in rock

DUBO0836 Expanded group in front of elephant in rock

DUBO0842 Group of men, maybe offering themselves to carry loads. Dubois describes such a group as wearing almost nothing, except for their bedlinnen, a reddish cloth, which they carried with them always. Shoes seem to be like the ones of the child in DUBO0801, possibly woven.

As there is no account of the last period of the Siwalik trip we can only presume. The trip must have been successful with regard to the main goal of Dubois, establishing a clear difference with the fauna and the geology of Java, thereby dismissing Martin's presumption about the presence of a Siwalik fauna in Indonesia.

But should Dubois have had any hopes of spectacular finds they were not fulfilled. He found lots of fossils, but nothing remotely humanoid or otherwise of earthshaking scientific importance.

DUBO1242 *Hemibos triquetricornis* RGM 3107, Siwalik, location Mahrawala

DUBO3277 Participants of the Third International Zoölogical Congress in Leiden 1895

## *The discussion about Pithecanthropus*

Dubois widely publicized his *Pithecanthropus* finds and after his return to Europe displayed them at several international congresses and scientific meetings although none of his early print publications hold pictures of the fossils nor of the finding sites and only very few particulars of the finding sites were the fossils had been discovered. The picture on the previous page shows the participants of the Third International Zoological Congress in Leiden 1895. Most of the players in this book were present at that congress, but only very few can be identified in this picture with certainty. Opinions on his discovery varied, but essentially the critics were divided into two camps. Some gave the *Pithecanthropus* remains the same treatment the Neanderthal fossils had received forty years earlier: they regarded them as primitive, though fully human. Others, however, ascribed the bones to an (upright-walking) ape. Dubois did not fail to exploit this difference of opinion. He pointed out that for some of his colleagues the fossils were apparently too primitive to be regarded as human, while for others they were too human-like to be assigned to an ape. Consequently, *Pithecanthropus* must have been something in between.

In general there was also doubt if the specimens, the femur, skull cap and the molar belonged to one individual. Leonce Manouvrier considered the femur from a *Homo sapiens*, the molar from a great Ape, a gibbon, or an orang utan, and the skull cap looks like a young gibbon. Further, the femur has the size of a normal human being. If the skull cap belonged to the same individual as the femur, according to Manouvrier, then the size of the individual was of a normal human being. However, a normal human being with such a small brain size must have been an idiot or a microcephalic. That is also what Lydekker and Cunningham thought; the skull cap was from a microcephalic idiot. That the bones were belonging to one individual Cunningham excluded.

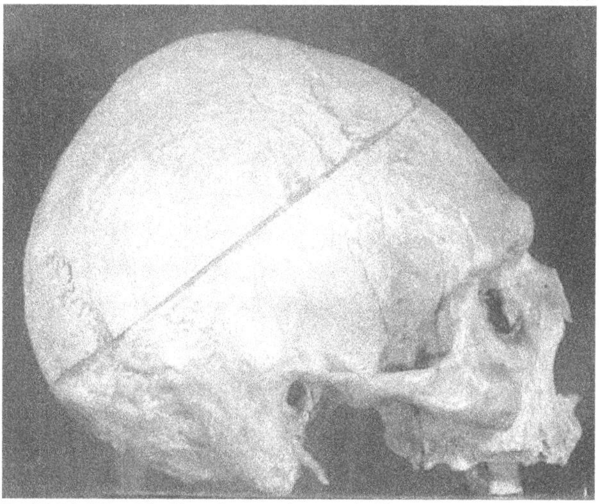

DUBO1459 Skull of a microcephalic man that Dubois acquired and used to point out to his opponents the differences between a microcephalic man and *Pithecanthropus erectus*.

DUBO4292 Rudolph Virchow

Rudolf Virchow, who had also discussed the Neanderthal skull, considered the skull cap of an Ape, and the femur human. Also Waldeyer considered the skull cap as belonging to an Ape and the femur as belonging to Man. But there were more scientists who considered *Pithecanthropus* as an Anthropoid. Representatives of those are Ramström, Boule and many others. Others considered *Pithecanthropus* a giant Gibbon, like Krause and Bumüller.

Turner and Topinard considered the molars to belong to large Orang utan. The most extreme interpretation is from Mahaudeau who thought *Pithecanthropus* to be a bastard from Man and Ape.

After Dubois had clarified several points, and especially after being on congresses, held during 1894-1900, where they had been able to examine the fossils for themselves, a growing number of scientists recognised that they were dealing with a transitional form linking humans with their ape-like ancestors. Thus, they accepted Dubois' belief that a phylogenetic significance could be ascribed to the fossils. Most of them, however, did not agree with Dubois' contention that *Pithecanthropus* stood exactly halfway between human and ape.

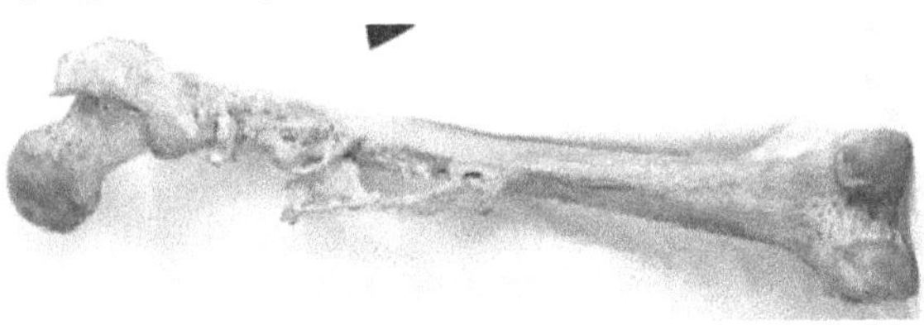

DUBO1312 Femur of a comedian who had exitocis compatable with the femur of *Pithecanthropus.*The man apparently did not suffer/notice this during his life and Dubois therefore used this example to rule out the argument of Virchow, who said that anyone with such exitocis must have been fully human because he must have had substantial social support to survive such disease.

In 1900 Dubois made a reconstruction of his transitional form for the World Exhibition of Paris informally referred to as 'Pete' ('Piet' in Dutch.)

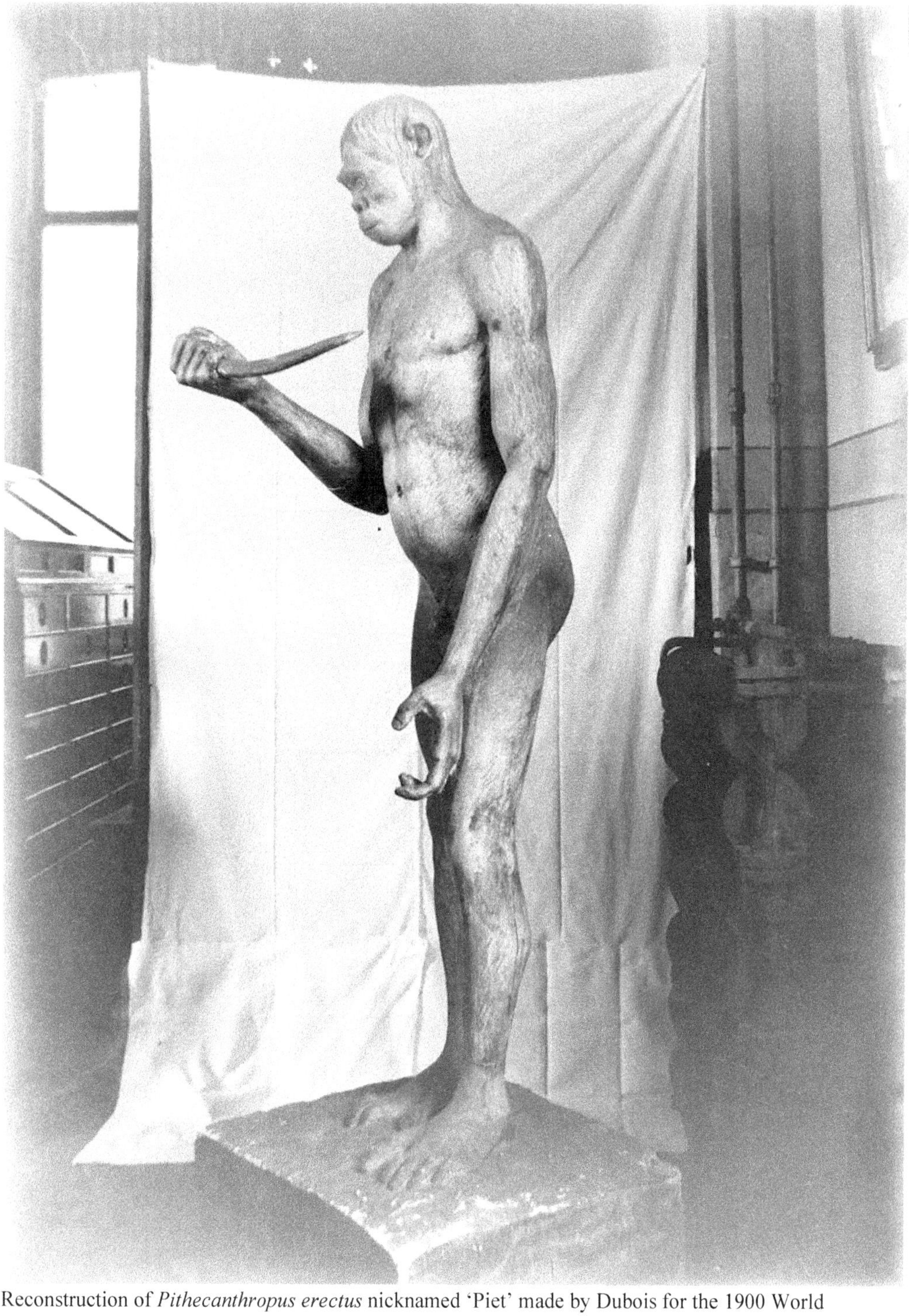

Reconstruction of *Pithecanthropus erectus* nicknamed 'Piet' made by Dubois for the 1900 World exhibition in Paris using his son as model. (Negative in Collection Geological Museum Artis, Amsterdam)

DUBO0001 Horse kept still by someone holding rein as the shadows reveal…

DUBO0548 Dogs are less easily kept still…(& Dubois' son Victor)

# Personal stuff

## *The letter*

When Eugène Dubois died in 1940 within the first year of World War II, his daughter Eugènie wrote to the Leo Brongersma, curator of Naturalis, the Museum of Natural History in Leiden, begging him to come and help sort out her father's affairs. Shortly before Dubois had definitely moved from Haarlem to his estate 'de Bedelaar' in the South of the Netherlands. The moving house however had been chaotic and lots of things had got mixed up. Eugenie was troubled by what awaited her. There was a war, her father's death, the chaos in the estate she inherited and the last will of her father which urged her to keep the estate which she could not afford. And now she had to beg Brongersma for help, although she knew that her father had fallen out with his former employee. She apologizes for her father's behaviour writing that Brongersma must surely know how mistrusting her father had always been to everybody and how he had always needed to have someone to blame for anything that went wrong. (Hooijer-Dubois, 1941). Brongersma had already become the curator of the Dubois collection in Naturalis, but now he also had been asked to go through Dubois' personal belongings to pick out those items that needed to be preserved for posterity. Alas, his view on what to keep and what not can in retrospect be regarded as regrettable (to say the least) and as a consequence after this first sifting many letters marked personal were burned. Whereas Eugènie may have done part of the selecting, and may have decided to burn the letters marked personal, a large part of the selection to be burnt has been made by Brongersma. He had not grasped that historians very much need that personal side as without it many aspects of an individuals' life just cannot be understood. We assume that the negatives of Dubois photographs have remained largely untouched, some for about a hundred years, but we do not know if any selection has taken place in these early years of the Dubois-collection. The fact that so few photographs of his personal life are among them makes one assume it might have.

Many, mostly glass negatives, were separately packed and numbered, and it seems only justified to start off with number one (left page), which is probably the most enigmatic photograph in a way, as we have no clue whatsoever why this horse has been photographed.

There are, except for a few family snapshots with dogs in it, no other photographs with just a live animal as a subject, which makes this one all the more peculiar. Maybe at some point someone will provide us with, what will no doubt be a clue of a personal nature, why this horse was kept still to have its picture taken.

DUBO2481(excerpt) Dubois and Eugènie at a meeting celebrating his 70th birthday (University of Amsterdam)

The letter of Eugènie played another role later when Bert Theunissen started his dissertation on the life of Dubois. Brongersma was not at all pleased that this 'young boy' started swimming in what he considered his private pool and he used the letter to legitimize his claim on the Dubois-collection and archives. As a consequence a substantial amount of data was kept hidden from the first biographer.

The second biographer, Pat Shipman acknowledged the important consequences of the letter in her preface, but did not mention the crucial role Brongersma had played and thereby suggests it was all due to Eugènie. We also tend to forget nowadays that burning of personal letters was probably a common practice, e.g. after the death of Charles Darwin the same happened with many of his personal letters.

## *Family*

DUBO0393 Anna Lojenga on boat?

Amongst the glass-negatives in the Dubois-collection the picture of a woman standing under what appears to be a sun cover on the deck of a ship is most likely the oldest one. It is a reasonable assumption that we are looking at Dubois' wife Anna Lojenga on the boat close to arriving in Indonesia, but the preservation of the negative is very poor and the facial features of the woman cannot really be made out. On the whole the quality of the negatives in the collection is rather good, but some, and particularly amongst the older ones, have suffered true time, faded, thumbed, and the photographic layer has sometimes come off. A few are broken, sometimes mended with sticky tape or have paper glued on them. Some of the family pictures are not on glass but on nitrate.

Of Anna's family we do not know a lot, but we do know she was the youngest of ten siblings, two brothers and seven sisters, the oldest being twenty years older than she was. Both her brothers served and died in Indonesia, as did her sister Margaretha Dingena, "Diete", who was married to a highly placed civil servant, Dirk Pieter Jentink, assistant resident on Borneo, present day Kalimantan.

Her mother, Willemien Jentink was related to the director of the National Museum of Natural History in Leiden, Frederik Anne Jentink, so all in all Dubois married into a family which had good contacts both in the scientific world and in Indonesia. Dubois used the contact to both Jentinks as becomes clear from a number of letters, but Diete dies already in 1887, shorly after the first letter exchanges which then cease entirely. F.A. Jentink keeps supporting and exchanging letters with Dubois extensively.

Jean, Victor & Eugènie Dubois in Indonesia (June 1892, collection N. Hooijer)

The lack of pictures of family members whilst in Indonesia has long suggested a quirky side of Dubois' character, that he regarded his family of much less interest than his fossils, but this needs to be corrected as it turns out family pictures did at some point exist: one is kindly provided by his great-grand-daughter depicting his children in June 1892 in Indonesia at the ages of 5, 4 and 1½.

After his return to the Netherlands at first a tour along family members appears to have been made (in a relatively short time judging by the age of the children) and furthermore a few portraits remain, of which some at the occasion of the holy communion of Eugenie, as well as a sea-side visit, some gardens and houses, a few indoor pictures, and two family x-ray pictures to end with. Most of them speak for themselves and will be presented without comment, names and places will be given when identified. Unidentified people are abbreviated as Woman A, Woman B, Child A etcetera, using the same abbreviations if the person(s) on different photographs appear identical.

DUBO0119 (nitrate) Eugenie, WA & WB

DUBO0120 (nitrate) WB, Eugenie & WA

For a number of years after they had returned to the Netherlands, Eugènie went to a private boarding school in Beverwijk where three of Anna's sisters were employed. WA and WB might well be two of these sisters with Eugènie in the school grounds.

DUBO0121(nitrate) Eugènie, WA & WB

DUBO0122(nitrate) WA & Eugènie

DUBO0123 (nitrate) Eugènie

The identification of grandma Lojenga is tentative and based on her being in traditional mourning clothes of the area in the Netherlands where Anna's parents originated from. Her father was a Protestant minister and had several subsequent constituencies, also in Elburg, where Anna was born. Mixed marriages between Roman Catholics (Dubois) and Protestants were not at all common in those days and in the Netherlands and in particular for a Protestant to marry a Catholic was considered degrading. Catholics were very much discriminated against and had much more difficulty in acquiring status and positions with political influence.

There is one more identification we want to suggest based on likeness: WD is very possibly yet another of Anna Lojenga's seven sisters and given that she seems to be of Anna's age it is possibly Willemien (Jr). Consequently, but less obvious in likeness, WC is likely to be her sister Ypke, who was married to Willem Dumortier, who according to the marriage certificate was witness to Dubois' marriage.

DUBO0559 Grandma Lojenga, Victor, Jean & Eugènie

DUBO0542 Grandma Lojenga, Victor, Jean & Eugènie

DUBO0545 Grandma Lojenga & WE

DUBO0544 Jean, Eugenie, Victor & Grandma Lojenga

DUBO0560 WG, Eugenie, WC, Victor, Jean

DUBO0553 Dubois'brother Victor and his wife, Marie Canoy, WA and her husband

DUBO0561 see pict 553 (left)

DUBO0546 WF, Eugenie, WC, Jean, Victor & WH

DUBO0549 CA, Jean, WC, Eugenie

DUBO0550 CA, Jean, WC, Eugenie

DUBO0547 Anna? & Victor

DUBO0552 Eugenie, CA, Jean, WC, Victor, CB

DUBO0551 Eugenie & Anna?

DUBO0554 Trinette? & grandchild

DUBO0543 Victor, WC, Eugenie, WD & Jean

DUBO0891 Gerardine and her kids

DUBO0993 Anna Lojenga

DUBO0983 Anna Lojenga

Dubois distanced himself from his Catholic background later in life (very probably because of vicious attacks from church officials because of his views on evolution and human descent) but was apparently embedded enough in the Catholic Church to have his daughter attend Holy Communion on the 10th of May 1898. For this ceremony the girls are dressed up like little brides. It is considered important enough to have his sister Mere Marie-Angelique be allowed to attend. Nuns were rarely allowed to visit their family except for catholic festivities like these and often did not visit their families for years...

DUBO0988 Mere Marie-Angelique

DUBO0985 Eugènie

DUBO0992 Eugènie

DUBO0894 Daughters of Gerardine?

DUBO0990 Janet Boyd(?)

DUBO0987 Willemien Lojenga-Jentink: mother of Anna, Victor, Anna Lojenga (dated 20-4-1898)

DUBO0989 Willemien Lojenga-Jentink: mother of Anna, Victor, Anna Lojenga CC, Jean (dated 20-4-1898)

DUBO0984 Eugènie, August 1909

DUBO0893 Victor, Eugènie, Jean

DUBO1300 Eugènie

The pictures of his children in the Naturalis collection represent roughly three phases: shortly after their return from Indonesia, a bit later on occasion of Eugènie's Holy Communion, and the last three, on this page, where Eugènie is adult.

It seems a bit odd nowadays that anyone should want to keep an x-ray picture of his wife's head or his daughter's hand, but it was not as macabre as it seems now. X-rays had only just been invented (Wilhelm Röntgen published his find first in 1895) and just a few years later anyone with money could get his or her head or hand x-rayed at a carnival or any other large gathering of marketers just for the fun of it. We asked a specialist if she could see any medical reason why these x-rays would have been made, but she could find no fault with the subjects. She remarked however that

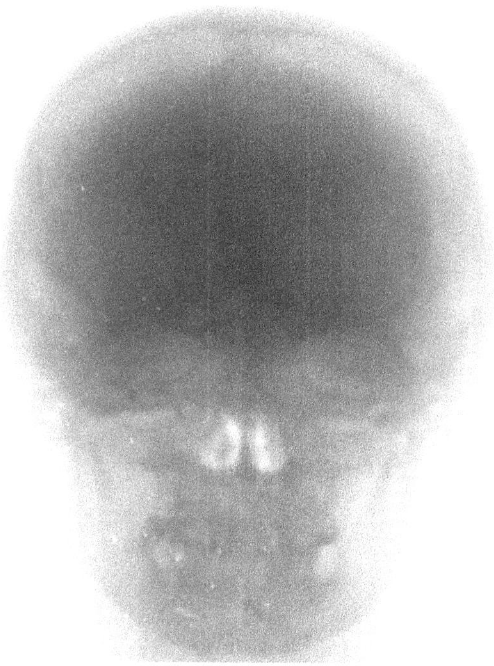

DUBO1515 X-ray of Anna Lojenga's head

the hand would not belong to the same body as the head, as it was clearly subadult, which can be seen by it having clearly visible growth-discs at the tarsal ends. The negative of the head carries the description 'Mrs. Dubois', so no doubt to whom that belongs. The hand carries two rings hence we describe it to the most obvious female subadult: his daughter Eugènie. As her age can be estimated relatively accurately at about eleven-twelve, the photograph can be dated at about 1899-1900.

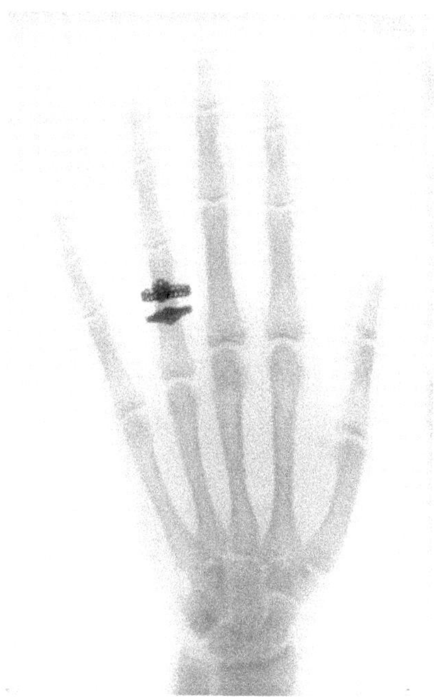

DUBO1325 X-ray of Eugènie Dubois' hand

DUBO0905 View possibly made out of Dubois'house at Sweelinkplein 12 The Hague at about 1896

DUBO1304 Formal garden in wintertime

DUBO0898 Garden see also DUBO0983 page 42

Dubois had an interest in gardens, at least in his later life he spent a lot of time (and money) in recreating in his garden at his estate 'de Bedelaar' an environment holding those plants that would have been around before the last Ice Age, the time where the fauna from Tegelen originates from. Some of the plants depicted are by all means extraordinary for that time in the Netherlands, the *Araucaria* centering DUBO0983 (page 42) with his wife rather in the background being the most explicit example, but also the *Oleander* in the background of the communion pictures would not have been a common plant in every household.

Two negatives are broken and both persons on them have not been identified. Sometimes the negatives have been written on as can be seen at the top of DUBO0557. It reads as Hoek or Holk, but it holds no clue to us to whom or where we are looking.

DUBO0557

DUBO0558

We intend to create a website via Naturalis on which we will log an up to date account of the pictures in this book, in hope you will be able to help us identify the people and places that we have not been able to identify yet. Do visit us at: http://science.naturalis.nl/dubois

First page with plates of the intended book on the Indonesian collection with Dubois' remarks in the sideline. Photo negatives remain in the collection.

# The book part I: The Indonesian collection

## *Understatements*

On December 15th 1913 Dubois writes a letter to the firm Brill, publishers in Leiden since 1683:

*Dear Sir,*

*although this year I have done the impossible to finish a first and most important part of my publication, which is to be published by you, on the Indonesian collection, I am[1] not been allowed to succeed. One circumstance that impeded me was the obligation to fulfill the wish of the Government, to receive a catalogue of the collection, which has taken much of the time available to me.*

*But also I could not have foreseen the size the field of aforementioned publication has amounted to over the past years. Much needs now to be examined from new points of view.*

*I request your trust, that I will do my utmost, to bring progress into this matter and I believe I can assure you that the long delay has not resulted in a decrease of interest. Interest now is surely larger than it has ever been.*

*In high esteem,*
*your servant,*
*Eug. Dubois*

This work would never finish. And when this letter was written it had already taken more than 15 years from the onset. We recently got access to the copybooks of the firm Brill in which all outgoing letters were copied and found a large number of letters[2] directed to Dubois that were hitherto unknown. From these letters it became soon clear that a contract had been made, which subsequently also turned up in the Brill archives. This very interesting document, handwritten in twofold by Dubois himself states that the Ministry of Colonial affairs will pay a sum of 6350 guilders for the completion of a book on the fossils of Indonesia, of which ƒ 2000,- can be claimed by Brill at the end of 1898, ƒ 2000,- can be claimed by Brill at the end of 1899 and the rest, ƒ 2350,- upon finishing of the whole lot.

Strangely enough there seems to be nothing in it for Dubois himself other than the honour of having the book in his name, but no financial incentive to him is stated. The Dubois archives hold a small booklet in which Dubois documented his expenses for Brill but as the ministry never got to see anything worthy of the money they reserved for it, Brill never got a penny and money becomes a serious issue in the relationship between Brill and Dubois. At that time, even ƒ 2000,- is a fair amount of money, equaling the annual income of a member of parliament at that time. So it is no wonder that at first Dubois is treated with the utmost respect and in particular Frans de Stoppelaar, one of the two managing directors at Brill, tries to endorse Dubois' every whim and even pays house visits to Dubois

---

[1] 'Am' instead of 'have' (which looks as odd in Dutch as it does in English) allows Dubois to express a personally felt responsibility (PA).

[2] All added in appendix I as is also the contract between Brill and Dubois.

to keep this project going. De Stoppelaar was involved in the organization of the Third International Zoological Congress in 1895 (he is probably one of the persons depicted at page 28) and must have noted the impact Dubois and his finds had on the scientific community. But that would all change and De Stoppelaar and later his companion Peltenburg would write many letters persuading, pursuing, arguing, and despairing but without coming anywhere closer to the goal they set themselves in 1897.

Meanwhile similar projects of Brill were very successful. For instance the famous Siboga expedition (1899-1900) to the Indonesian archipelago headed by the biologist Max Weber, resulted in a series of no less than 142 publications, well into the nineteen seventies. But there are also reports on expeditions to New Guinea and Borneo and even reports on American expeditions to Alaska and Siberia that were successfully published by Brill.[3]

Max Weber, headed the Siboga expedition and supported Dubois on many occasions

However, to anyone who knew the extent of material Dubois had shipped to the Netherlands it must have been quite clear from the start that Dubois underestimated the amount of work needed. But furthermore, he was not the man to fiddle about with details, as having to deal with details slowed him down. If the Government had not insisted on a detailed catalogue of the 45000+ items he had put in the museum, the whole collection would now have been of little or no value. Lots of information necessary to understand the fossils and to interpret them for research would have been lost and taken to the grave by him. He understood the importance but if he had not been forced to accept assistance the catalogue would have suffered the same fate that the book did, and by the time the letter to Brill was written in 1913, the catalogue would still take more than 20 years to finish. But it did finish (solely through the work of his assistant, the Jesuit father Sanctes Bernsen) and the whole collection is therefore still of great importance today. As an example, his great nemesis of later days, Ralph von Koenigswald, who also collected a large amount of fossils in Indonesia, which are now in the Senckenberg Museum in Frankfurt, did not provide any decent catalogue to his finds. As a consequence many are now in essence lost for science as much of the necessary information on the location and circumstances of these finds has been lost. In fact only by comparison to the Dubois collection some of this lost information is now being retrieved (by the diligent work of Christine Hertler), a fact Dubois will surely be gloating about to Von Koenigswald should they both have been wrong about there being no afterlife…

[3] Brill has printed a book celebrating its 325th anniversary which is the source of many details like these in this text; the book is, however, not commercially available. (ISBN 978 90 04 17031 5)

## *Contents*

The intended contents of the book to be can only be partially reconstructed. There is just one notebook that seems to contain manuscripts that might have been first drafts of ideas and a small booklet with texts accompanying the photos made in Indonesia. Next to that there are the plates that were printed by Brill, some printed proofs of text and a number of photographs that seem to be made for publication but never used. The printed proofs of text are not really interesting, they are repetitions of his work on the *Pithecanthropus* material and his ideas on the relation between brain volume and body size (This subject is treated extensively by Bert Theunissen).

The notebook states on the first page that it contains notes and texts for the manuscript on the fossils from Java, but the actual contents are disappointingly dull and do not add anything useful to what is already on the printed pages.

The plates however only partly match the text, numbers 1-13 covering sites where Dubois surveyed and 14-24 covering *Pithecanthropus* and material used in comparison. But it is seems only logical that also the species that had been found should have been treated. Dubois made a small start and it seems he wanted to start with the lower land vertebrates: there are some printed skulls of crocodiles and gavials. We will just start with a selection from the pictures of the species that he brought home with him. Subsequently we will deal with some of the pictures we have of sites, part of which are on the plates that Brill had printed and part of which we only have negatives/printed photographs. Information belonging to these sites is generally sparse.

We will end with the material of *Pithecanthropus* and the material to which it was compared and used as subject on the plates printed by Brill.

## *Species*

Dubois considered the fossil remains found by him in the various localities together to represent faunas of similar age to which he applied the name Kendeng or Trinil fauna. Later in the early 80ties of the last century the so-called Trinil fauna of Dubois was split up into the Trinil fauna sensu stricto and the Kedung Brubus fauna, based on differences in faunal composition. The two faunas were considered to be of different ages: Kedung Brubus with an age of 800.000 younger than the Trinil fauna sensu stricto with an age of one million years.

One of the problems Dubois had was that he did not have the space available to lay out his collection in such a way that he could get a good overview. Surely Dubois wanted to describe and figure the fossils himself for the book he had planned. But although photographs were made, they seem only preliminary. The pictures we have selected in this section are only from those of which we are certain that they are early ones made or commissioned by Dubois himself. There are more pictures in the collection which probably and sometimes certainly have been made by Brongersma or Hooijer, and alas not always can be distinguished anymore from the original Dubois collection, but of the ones presented here we are ‘pretty sure’...

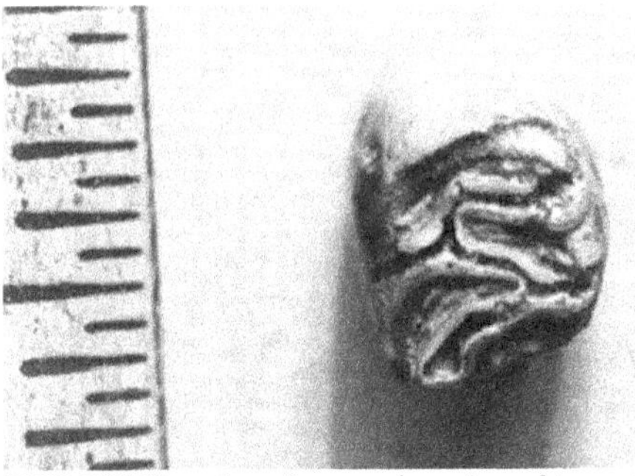

DUBO0279 molar of porcupine *Acanthion brachyurus* LINNAEUS 1758, a porcupine. Only a few specimens were found at Trinil.

DUBO0823 upper jaw of *Rhinoceros sondaicus* DESMAREST, the Indian rhino. This species is represented in the Dubois collection by various specimens.

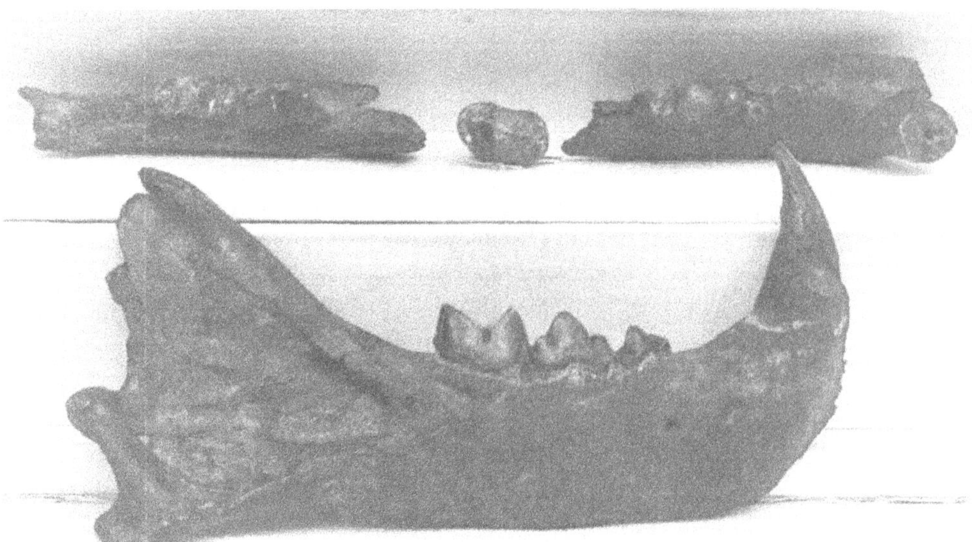

DUBO0928 Lower specimen hemi mandible of *Panthera trinilensis*
*Panthera tigris* LINNAEUS, 1758, the tiger. According to Dubois there were two species of fossil tigers in the collection; the one from Trinil, he called *Panthera (Felis) trinilensis* and one from Kedung Brubus, he called *Panthera (Felis) oxygnatha.* Later they became synonym for *Panthera tigris*, however as extinct subspecies: *Panthera tigris trinilensis* and *Panthera trigris oxygnatha.*

DUBO0936 *Stegodon trigonocephalus* MARTIN mandibles (top) DUBO0937 molars (bottom)
*Stegodon* is an extinct proboscidean (elephant like creature) of which the molars are low crowned. Remarkable of this species is that the tusks are not spreading, but lie so close together that the trunk cannot fit in between anymore  A lot of specimens from this species are found.

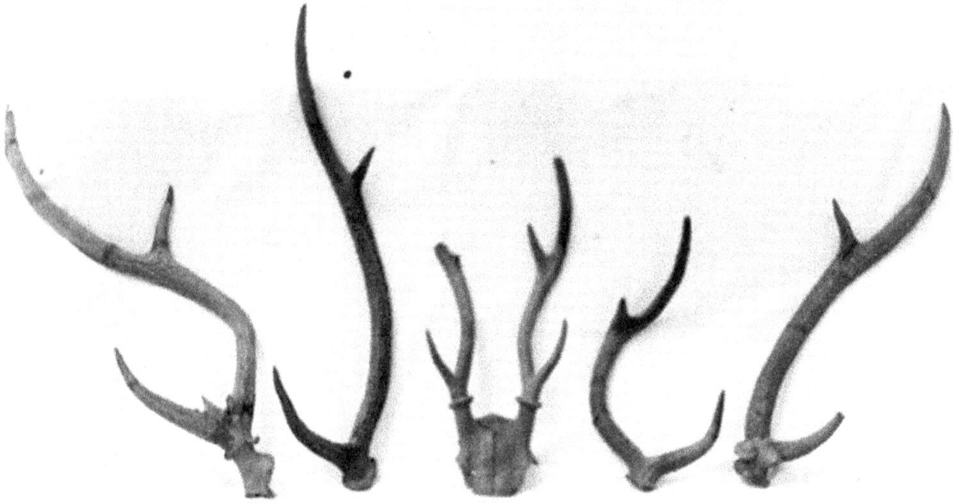

DUBO0977 Skull (middle) and antlers of *Cervus (Axis) lydekkeri* MARTIN
Hundreds of antlers, mandibles etc. of this extinct species are present in the Dubois collection. Dubois named it *Cervus liriocerus*, but it is now a synonym of *Cervus (Axis) lydekkeri* MARTIN

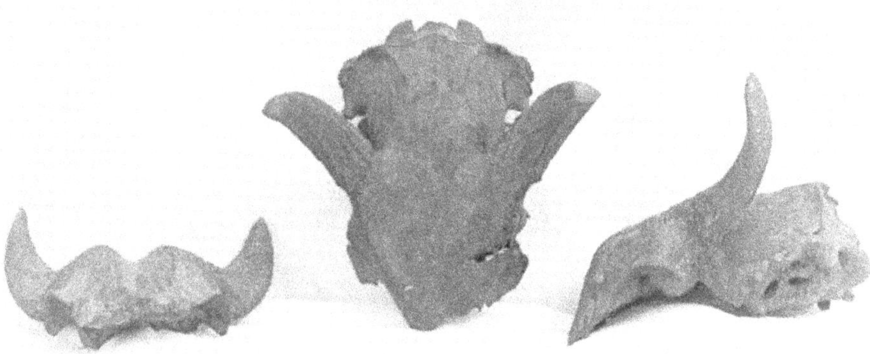

DUBO0981. Skulls and horn core of *Duboisia santeng* DUBOIS, 1891, extinct *Boselaphide*, related to the Nylgau.First this rather small extinct species, with the size of a goat, was named *Anoa santeng* by Dubois, later he named it in honour of Kroesen, who was the Governor of the Dutch East Indies, *Tetraceros kroesenii*. On the label is written: Boselaphus Kroesenii van Trinil (af. Kgawi, Res. Madioen, Java) This species is only known from Java.

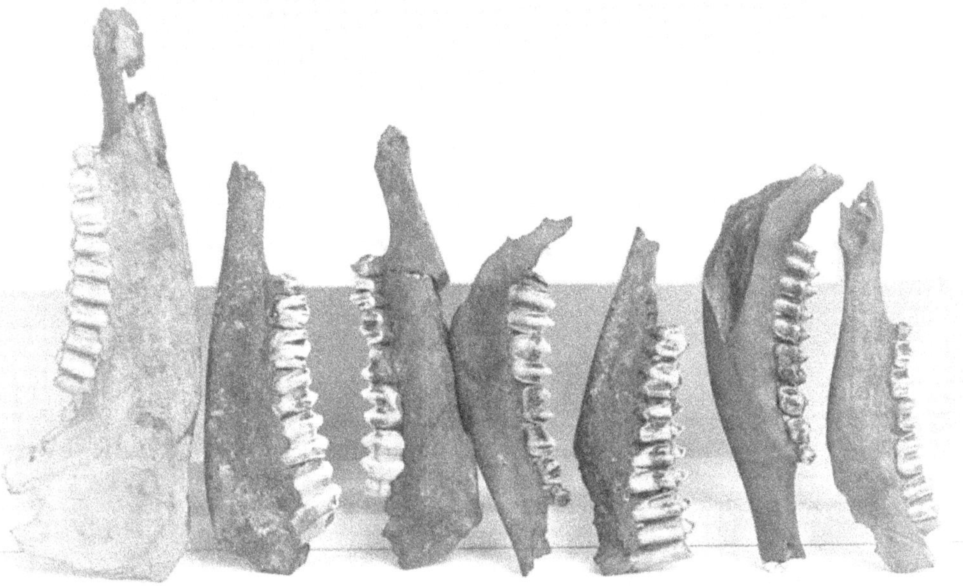

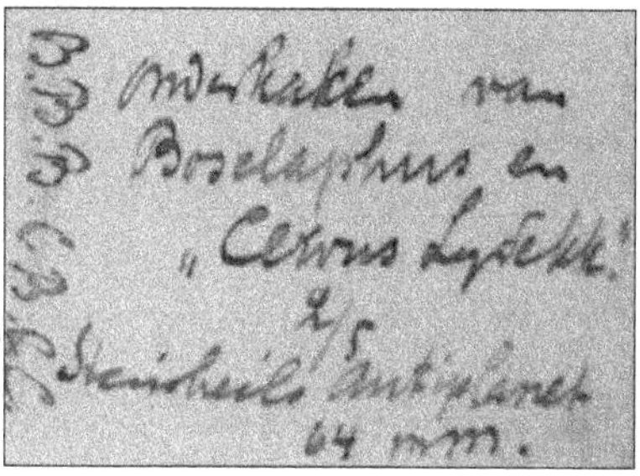


DUBO0938 mandibles from *Duboisia* and *Axis lydekkeri*. *Duboisia santeng* and *Axis lydekkeri* are of similar size. Dubois pictured the mandibles of those two species in one photograph. Original label superimposed.

DUBO0978 skull and horn core of *Bibos palaesondaicus*, 1908, an extinct banteng. This fossil banteng has oval horn cores in cross section.

DUBO0976 Skull and horn cores of *Bubalus palaeokerabau* DUBOIS, 1908, extinct water buffalo. This extinct species is very well represented. The horn cores can be easily determined, as the outline of their cross section is trianglar in shape. The horn cores can have a width from tip to tip of more than 2 and a half meter. Thousands of molars have been found of the bovid species *Bubalus palaeokerabau* and *Bibos palaesondaicus*, which are difficult to separate.

DUBO0923 Upper row: molars of *Acanthion brachyurus*; middle and lower row: all the specimens of monkeys, *Macaca* sp./ *Trachpithecus cristatus robustus* HOOIJER, 1962, in the Dubois collection, with the exception of the most left and most right molars of the middle row, which are from *Panthera tigris*.

Whereas afore mentioned species are from the Trinil site, the following are from Kedung Brubus and not present in Trinil sensu strictu.

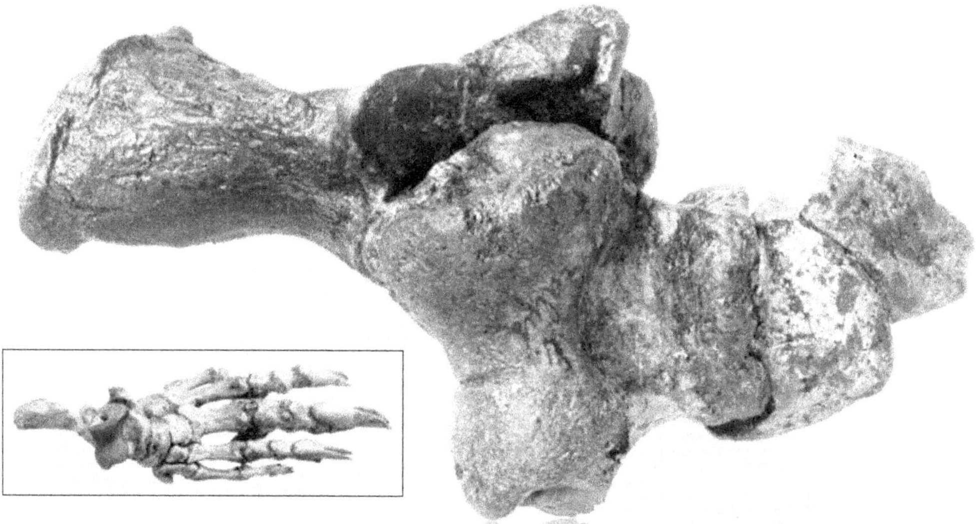

DUBO0958. Articulated foot bones of *Manis palaeojavanica* DUBOIS, 1907: the scaly anteater. From this extinct species material from one animal was found. This is remarkable, as it is the only skeletal material in association and is also the only species which was extensively described by Dubois in 1926. It has twice the size of the recent scaly anteater *Manis javanica*. Insert: DUBO0959 complete foot of a recent *Manis* for comparison of shape (not size)

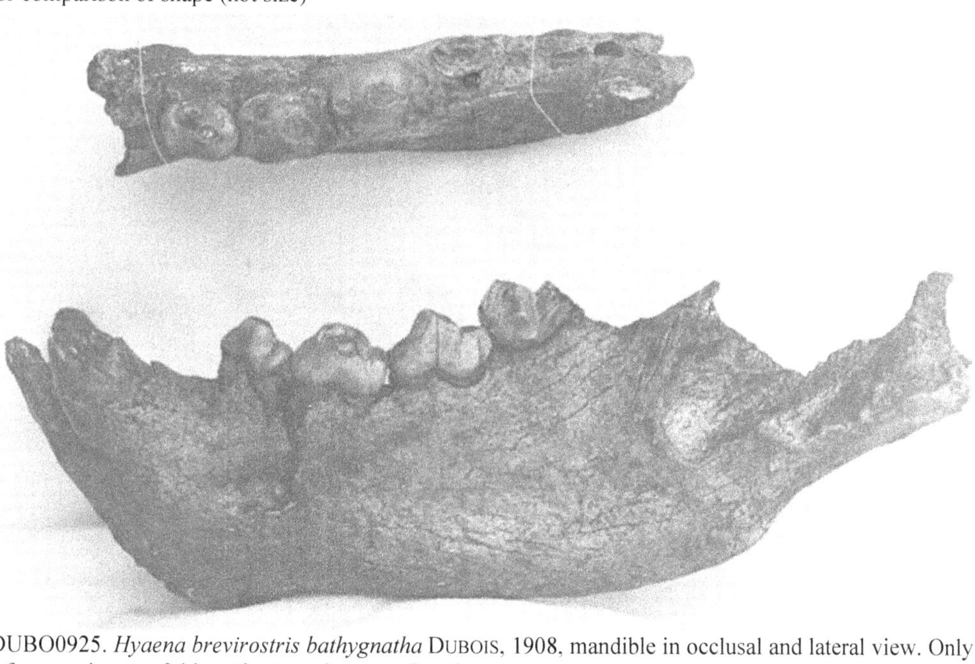

DUBO0925. *Hyaena brevirostris bathygnatha* DUBOIS, 1908, mandible in occlusal and lateral view. Only a few specimens of this extinct species were found.

DUBO0940. *Duboisia* maxilla's, with, in the middle, lower row, part of maxilla from *Lutra palaeoleptonyx* DUBOIS, 1907, an otter. From this species only one specimen was found. Inserted: part of the original label.

DUBO0945 *Elephas hysudrindicus* DUBOIS, 1908, molars. Although thousands of fossils were found in Trinil, this species is not present in Trinil sensu stricto. Based on this the fauna from Kedung Brubus is considered to be younger.

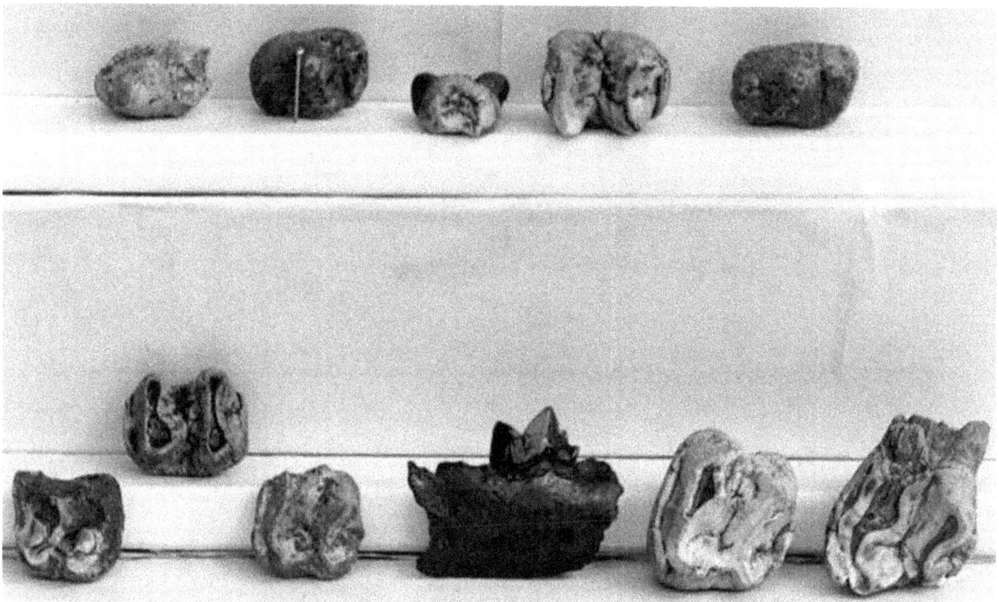

DUBO0929. *Tapirus indicus* DESMAREST molars: Upper row, 4th of left side a lower molar; in the lower upper and lower molars, with the exception of the 4th of the left which is a molar of tiger. This species was described by Dubois as *Tapirus pandanicus*, but Hooijer considered the specimens to be from *Tapirus indicus*. In the photograph all the specimens of *Tapirus* which are found are present.

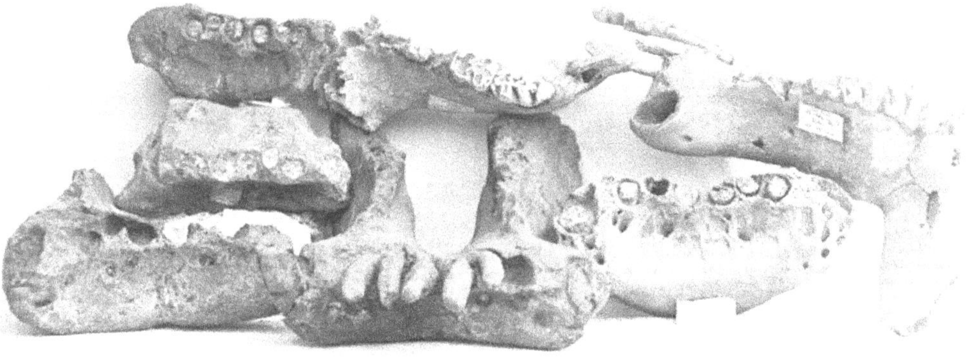

DUBO0822. Mandibles of *Hexaprotodon sivajavanicus* DUBOIS, 1908. *Hexaprotodon* is an extinct hippopotamus. It differs from the recent hippopotamus (*Hippopotamus amphibius*) in having 6 incisors, instead of 4 (see the middle mandible). Hooijer considered the hippopotamus to be the same as the one from the Siwaliks: *Hexaprotodon sivalensis* FALCONER AND CAUTLEY. Dubois named the species *Hippopotamus (Hexaprotodon) sivajavanicus*, which is its name again today. DUBO0830 right, shows a partial maxilla of *Hexaprotodon*

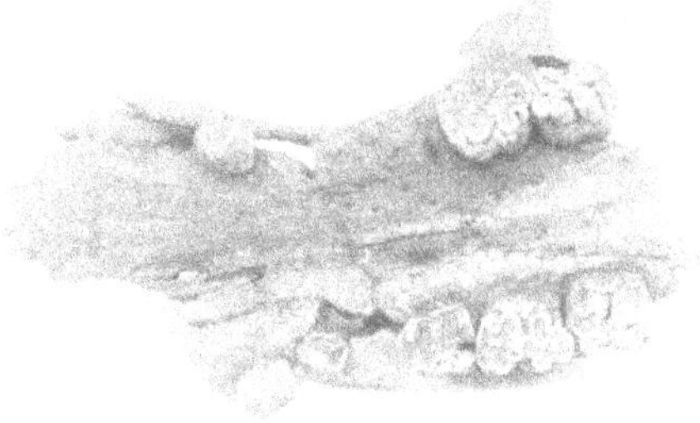

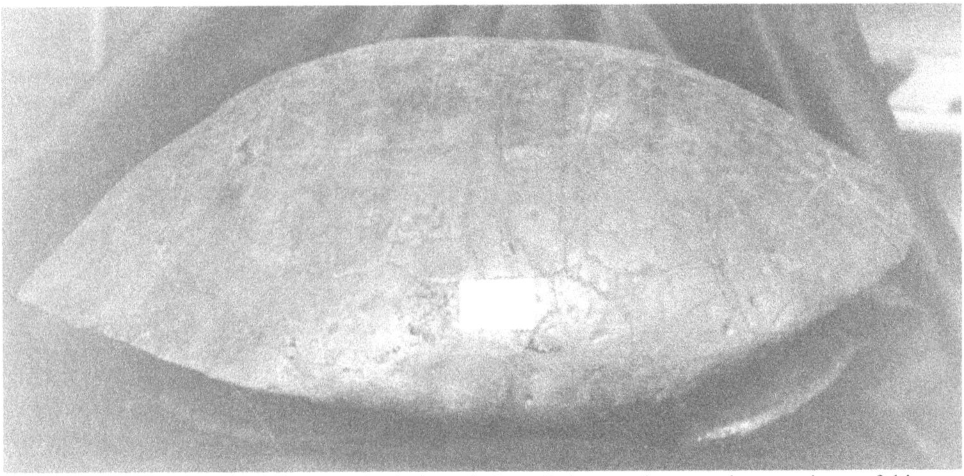

DUBO01022. *Isoclina hardella* a tortoise in lateral view. Only one nice complete specimen of this species is present.

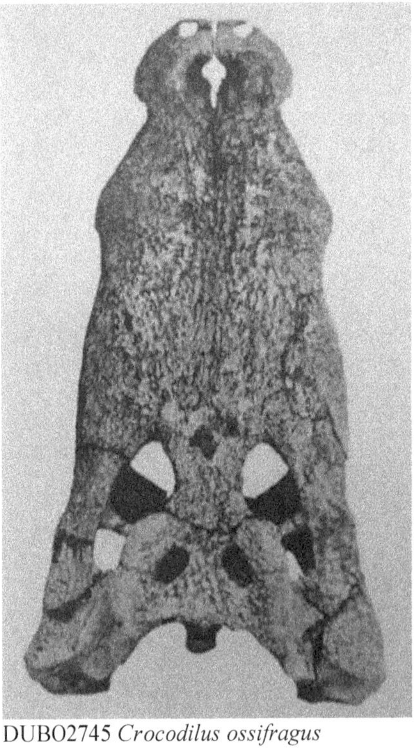

DUBO2745 *Crocodilus ossifragus*

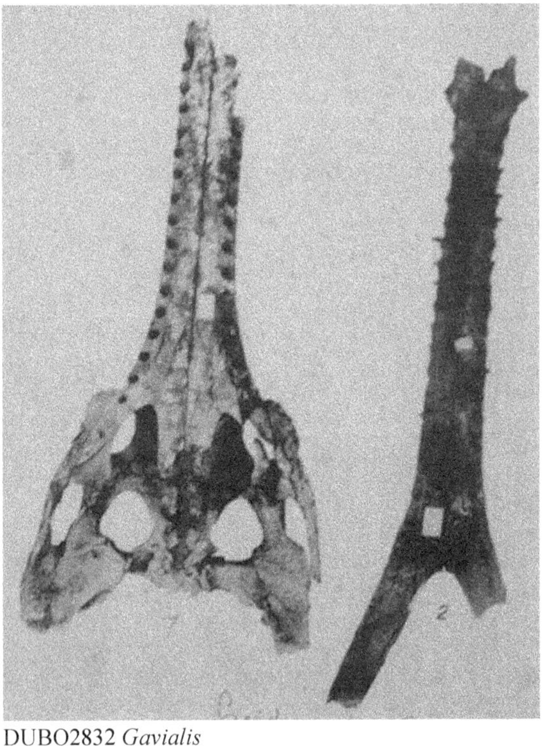

DUBO2832 *Gavialis*

Besides the mammal species mentioned so far, also a lot of specimens from crocodiles, gavials, tortoises, birds and molluscs are found. In particular a lot of material from crocodiles is present and at some point this material was considered for publication, most likely for the book, as we have a few proof-like prints from which the two figures below here are examples.

Several species have not been depicted here because they either were not present in the Dubois collection at all or, if they are present, we think they are post 1920, hence not made with the intention to be in the book. Of the Trinil species these are e.g. *Prionailures bengalensis* (Kerr) a small cat: from this species only one specimen, a mandible, was found in the trench made in 1900 and was indicated by Dubois as *Felis microgale*. *Sus brachygnathus* DUBOIS, 1908; an extinct pig. The pig with the short face, is reasonably represented in the Dubois Collection. *Muntiacus* sp. This species is represented only by a few specimens. And finally there is also a rat: *Rattus trinilensis* MUSSER, 1982.

The Kedung Brubus fauna also holds more specimens of *Panthera tigris* LINNAEUS, 1758, the tiger. The species from Kedung Brubus was called *Felis oxygnatha*, but turned out to be an extinct subspecies of *Panthera tigris*: *Panthera tigris oxygnatha.* Next to that there are also a rhinoceros, *Rhinoceros kendengindicus*, DUBOIS, 1908 and an extinct pig, *Sus macrognathus* DUBOIS. This species is larger than *Sus brachygnathus*. Again the deer like *Muntiacus* sp. and an extinct bovid, *Epileptobos groeneveldtii* DUBOIS, 1908, depicted on page 121. Compared to *Bubalus* and *Bibos* this species is rare. The cross section of the horn cores is round. It was named by Dubois after Willem Groeneveldt, director of the Department of Education, Religion and Industry, who supported Dubois and was responsible for the scientific research in the Indies.

DUBO0504 *Eamesiella aff. corrugata* DESHAYES

The molluscs he collected require special attention. Dubois fully acknowledged their importance in establishing the age of the layers he had collected them from but lacks the knowledge to make any sense of this particular subject. In 1928 he is being sent a reprint of the work of Tera van Benthem Jutting on the "Marine molluscs of the island of

Curaçao". He immediately writes to her to congratulate her with her great achievement and asks her if she would not also be willing to work on the Trinil and Kendeng molluscs.

In this letter later in his life it is particularly interesting to note that Dubois is not being paranoid about his finds at all, he freely gives up his material for research and explicitly states that the work be published under her name and not his. His paranoia seems very much only to be limited to his *Pithecanthropus*, he equally easily has parted from the carnivore fossils in favour of Brongersma, and the Tegelen-fauna in favour of Bernsen and Schreuder.

All finds of Dubois' expedition are still of great importance and help reconstructing the environment in which *Homo* (*Pithecanthropus*) *erectus* must have lived. Based on the large quantity of fossils of bovids, *Stegodon* and rhinoceros, the environment must have been savanna-like. The occasional tree would have provided shelter for the odd monkey and a river must have crossed the savannah, where the crocodiles and the hippos would have found their repose. The mollusks were described by van Benthem Jutting in a booklet published by Brill in 1937: "Non marine mollusca from fossil horizons in Java with special reference to the Trinil fauna." And even the interest in these mollusks has still not tanned as proved by the publication of Joordens in 2009: Joordens, J.C.A., et al., Relevance of aquatic environments for hominins: a case study from Trinil (Java, Indonesia), J Hum Evol (2009), doi:10.1016/j.jhevol.2009.06.003

DUBO0500 *Pseudodon vondembuschianus trinilensis* DUBOIS

## *Locations*

Next we present most of the pictures that Dubois had made of locations in Indonesia. Many were surely intended for future publications. The plates were intended for his book.

DUBO1101

DUBO1100 Location (marked with little stick) where a red painted skull (skeleton) was found in front of the hoekgrot site (entrance of the cave left).(1890)

DUBO1101/5/6/7/8/9 are all views on the same ridge where the cave is located described by Dubois as views on Gunung Låwå near Tjermee as seen from the south edge of the dessa Tjampoer. On the edge of the negative of DUBO1106 the numbers indicate 1) The west cave (with human remains)

2) Cave in the steep rock (contains just tenacious låwå)

3) Location where the Rietschoten-skull has been found

These sites were extensively treated in the dissertation of Paul Storm: 'The evolutionary significance of the Wajak skulls'. Scripta Geologica 110 (1995)

DUBO1105

DUBO1106

DUBO1109

DUBO1107

DUBO1108

DUBO1102 and DUBO1103/4 on the next page are all views from Hoekgrot site looking down, of no particular scientific interest, but no doubt for the photographer to good an opportunity for a landscape photograph to let pass…

DUBO1103

DUBO1104

DUBO1272 Ravine of the Kali Brubus about 1 km N.E. of the Kendeng Brubus, sloping layers

DUBO0707 Tuff wall 100 m N.E. of Kedung Brubus (ED)

DUBO1263 Tuff wall at Kali Ngeto (Kedung Ampel) 470 m N.E. (along the direction of the slope) of the wall B by Kedung Brubus N.E.-N > Sandstone above Claystone > S.W.-S, Stick = 1 m (ED)

DUBO0710 Current bedding seen in the direction of where the river along Kedung Brubus used to be. E. of Ked. Brubus. Wall is N.E.>S.W. Perpendicular on the direction of the river. Stick with ½ m. (ED)

DUBO1308(left) and DUBO1289 (right) Tanah cleft, 12 m high wall at a bit more than 1½ km N.W. of Bangle. Claystone beneath a stony tuff. Claystone with layers of hardened chalk. Contains freshwater shells known from elsewhere and pieces of bone NE > SW (ED) (not all words could be deciphered…)

DUBO1259 Tanah cleft, Bangle, sloping of tuff layers.(ED)

DUBO1283 Top of plate 1. Andesite tuff holding bones overlying Pliocene marls along the Solo river(ED)

DUBO1257 Bottom of plate 1. Tuff wall bordering Kali Brubus about 200 m SE of crossing towards Kedung Madoh, slightly beneath the left little ravine. Sandstone is sloping from N. N.E. towards S. S.W. and the current bedding is N.E.-S.W. Little holes are made by crabs (ED)

DUBO1274 Top of plate 2. Tuff wall about 150 m N.E. of Kedung Brubus, along the Kali Ngeto. Yellowish sandstone 8 m to the top, amber grey clayloam 5 m until the little river. The man is about 1.65 m. (ED)

DUBO1269 Bottom of plate 2. Sand-claystone wall close to the mouth of the Kali Ngeto near Kedung Brubus. Current bedding N. N.W. Sitting man is 0.90 m high. The horizontal line behind his neck is the border with the claystone (ED)

DUBO0709 This photograph is very much similar to the top of plate 2, maybe taken at a different moment in time, more probably at a close by but different location. This time the man is standing halfway the height, probably indicating with his feet the border between clay and sandstone...

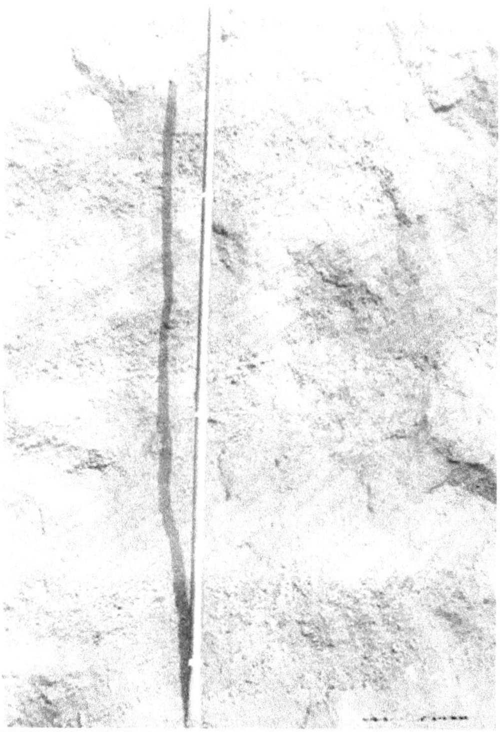

DUBO1137 Left on plate 3. Tuff wall along the Kali Ngeto near Kedung Brubus with beautiful diagonal banding. S. S.E > N. N.W. The band with the diagonal layers is parallel to the border with the clay stone. Stick with ½ m marks. (ED)

DUBO1175 Right on plate 3. Current bedding in wall near Kedung Brubus. (N. of the total view) S. S.E > N. N.W.

DUBO1254 Top of plate 4. Small hill of 'bone'tuff near Boetak. (About 380 m S. down) View from the N.E. Slope 10,5° N. N.E. > S. S.W. 6-12 m height. Sandstone on top. Clay/loam below. Beneath this an andesite breccia, invisible on the photograph (ED)

DUBO1278 Bottom of plate 4. Small hill near Boetak. View from S.E.

The small hill near Butak is a geological oddity in the landscape that was the subject of some discussion but was not of great significance to the fossil collection. Dubois however gave it considerable attention as there are 8 different negatives of the same location. The difference in dress suggests two different dates for these photographs. The one man is on most photographs openly wearing a klewang, the Indonesian machete, which was also a useful weapon. This could have had some political significance that Dubois only understood after the pictures had been made and thus forced him to make new ones with someone more neatly dressed. Another reason may have been his naked torso, but neither reason would have mattered much to Dubois himself, as long as he knew the man's length. They were both 1,65 m. (Maybe it is the same man after all)

| DUBO1267 | DUBO1267 | DUBO1268 |
|---|---|---|
| DUBO1266 | | DUBO1265 |
| DUBO1255 | DUBO1282 | DUBO1282 |

DUBO1279/87 Top of plate 5. As the quality of these negatives is poor, we used the (retouched?) print here instead. Maybe the used negative is lost. — Current bedding in sandstone of small hill near Boetak. S.E. > N.W. Lines of wear by roots. (ED)

DUBO0562 Bottom of plate 5. Current bedding near Kebon (along the Bengawan) District Sapteh Ngawi (ED)

The fact that Dubois notes all these current beddings and stratigraphies indicates that he is very much aware of the importance of these data for the interpretation of his material. The diagonals within the horizontal bands indicate the ancient direction of the water flow of the river. The geological work of Dubois, however, has not received much attention yet.

Plate 6. Railway cut about 3 km W. of Wilangang station. Total length about 500 m. Beautiful layers of tuf and rolled stones. Top — View from near the east end westward. Bottom — view from the near the middle eastward (ED) Corresponding negatives (DUBO1135/36) are of poor quality and probably not original (photocopied of a print)

DUBO1133 Plate 7. Railway cut about 3 km W. of Wilangang station. Near the middle. Stick length ½ m. (ED)

DUBO0690 Plate 8. The Trinil site along the Solo-river on Sept 5th 1894. View from near the monument.

DUBO1494 Plate 9. Overview of the Trinil site along the Solo-river. Photograph made by a photographer arranged by Kriele on Dubois' request at 19-11-1900. Again made from near the monument. This perspective is the same as of the photograph used for the cover (DUBO1402).

DUBO1399 Plate 10. Overview of the Trinil site along the Solo-river from yet another perspective (1900).

DUBO1400 Plate 11. Last overview of the Trinil site along the Solo-river.

DUBO1134 Plate 12. Monument that Dubois had erected on September 5th 1894.

DUBO1273 Plate 13. Overview over the Kendeng (ED) This is the last plate that Dubois intended for his book as far as the descriptions of locations are concerned. The white line is the retouch accentuating the skyline of the mountain range. The printed version (below) is much enhanced. The significance of this particular view is not clear: it may well express some sentimental touch.

## *Pithecanthropus and its comparative material*

Whereas for the previous material finding the correct negative with plates for the book is relatively easy, for the skull material this is not the case (nor is it very relevant). Dubois photographed the key material dozens of times with ever so slight variations in position or lighting. Whenever a DUBO source number is mentioned you need to keep in mind that after close examination we might not have come up with the exact fit but just something very close. The prints are more relevant here than the photographs, and have been extensively retouched and commented on by Dubois before being printed in their final stage. The prints are what he considered a truthful representation, more so than the photographs themselves. We will therefore show the plates rather than the numerous photographs.

Eventually, as we will describe later, 30 years after Dubois found the fossils, he finally publishes photographs with descriptions. The descriptions given on the plates in boxes are derived from those publications.

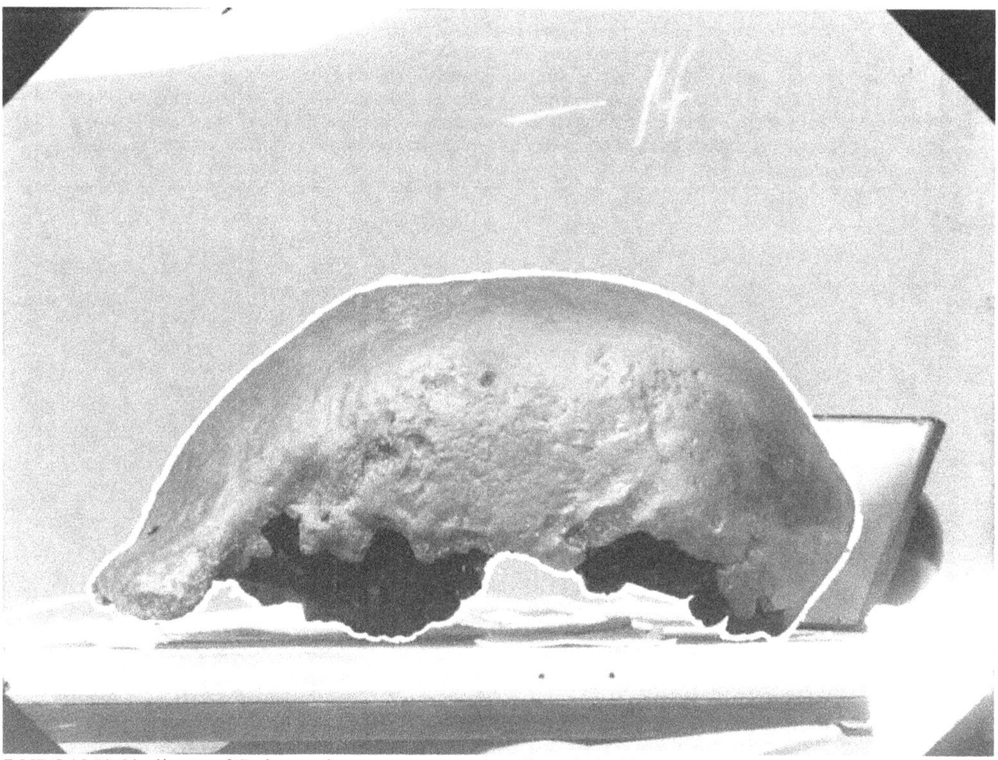

DUBO1359 Skull cap of *Pithecanthropus erectus*, view from left side. This photograph seems to be the one used for the bottom of plate 14. The white line is retouch to accentuate the outline for printing purposes.

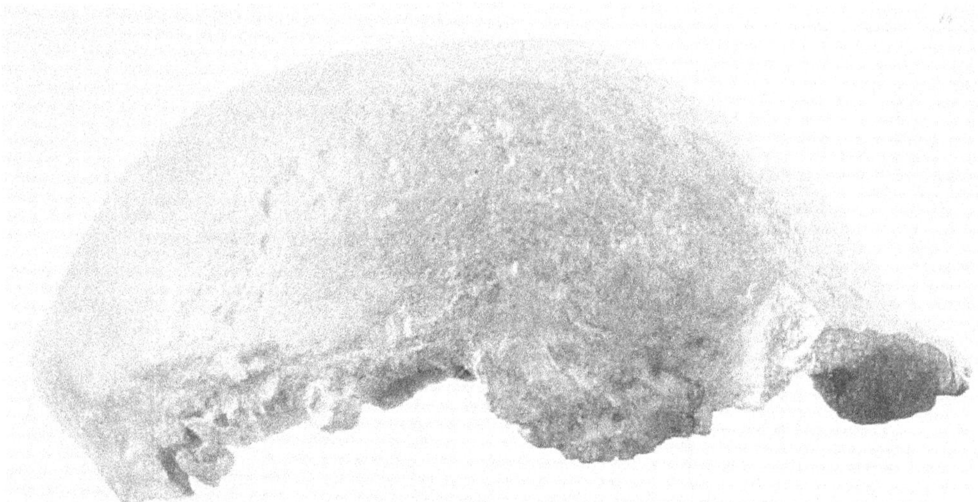

Calvarium from the right side. The loss of bony substance is not as large as on the left. The fracture surface of the external angular process is very clear. The deepest part of the temporal fossa 45 to 50 mm. from the glabella. Sharp bending of the parietal surface, at about the place of the temporal line, chiefly due to the loss of the outer table of the bone on the side of the sagittal suture. On the border of the calvarial fragment, 6 cm before the torus occipitalis, a very compact but externally much corroded part of the petrosum, 1 $cm^2$ in extent, immediately before which the place of the external auditory meatus must have been. At and near the occipito-mastoid suture furrows and parts of canals for the blood vessels: the foramen mastoideum on the border line. On comparing the external outline of the borderwards increasingly defective nuchal part of the occipital bone, in these figures, with the other figures, it becomes evident that the angle of the nuchal part is not human but pithecoid.

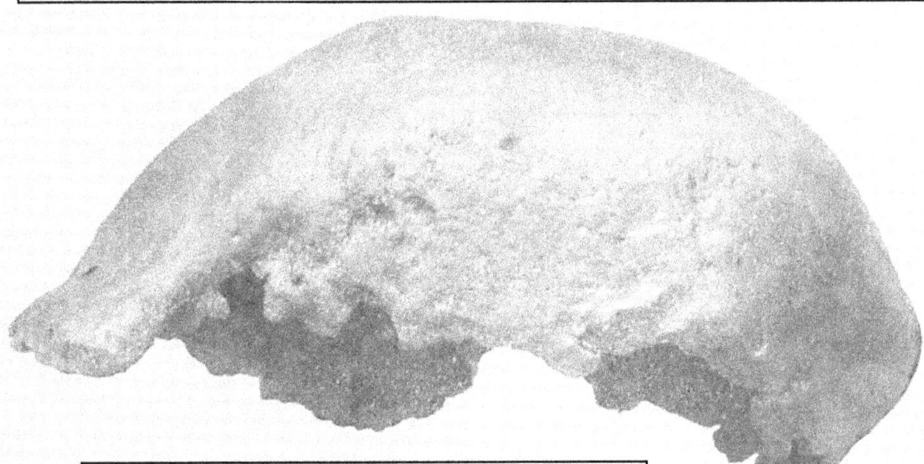

Calvarium from the left side. Sagittal (median) glabella-inion plane. This side, especially in the temporal fossa, deeply corroded, in some places even to the vitreous table, at the border even to perforation.

Plate 14 (Top DUBO1362, Bottom DUBO1359) *Pithecanthropus erectus*

Plate 15
(DUBO1358)
*P. erectus*

Calvarium from above.Transversal glabella-inion plane. Extensive corrosion of the surface. The white-spotted areas behind the coronal suture, on both sides of the sagittal suture, are places where the outer wall of the parietal bones has entirely disappeared and the diploë laid bare. The same at the much worn off left in an oblique line. The natural constriction of the frontal is only conserved on the right side. Evident trigonocephalism. The obliquity of the occiput is due to corrosion. The smooth area on the hinder part of the parietal bones is artificial.

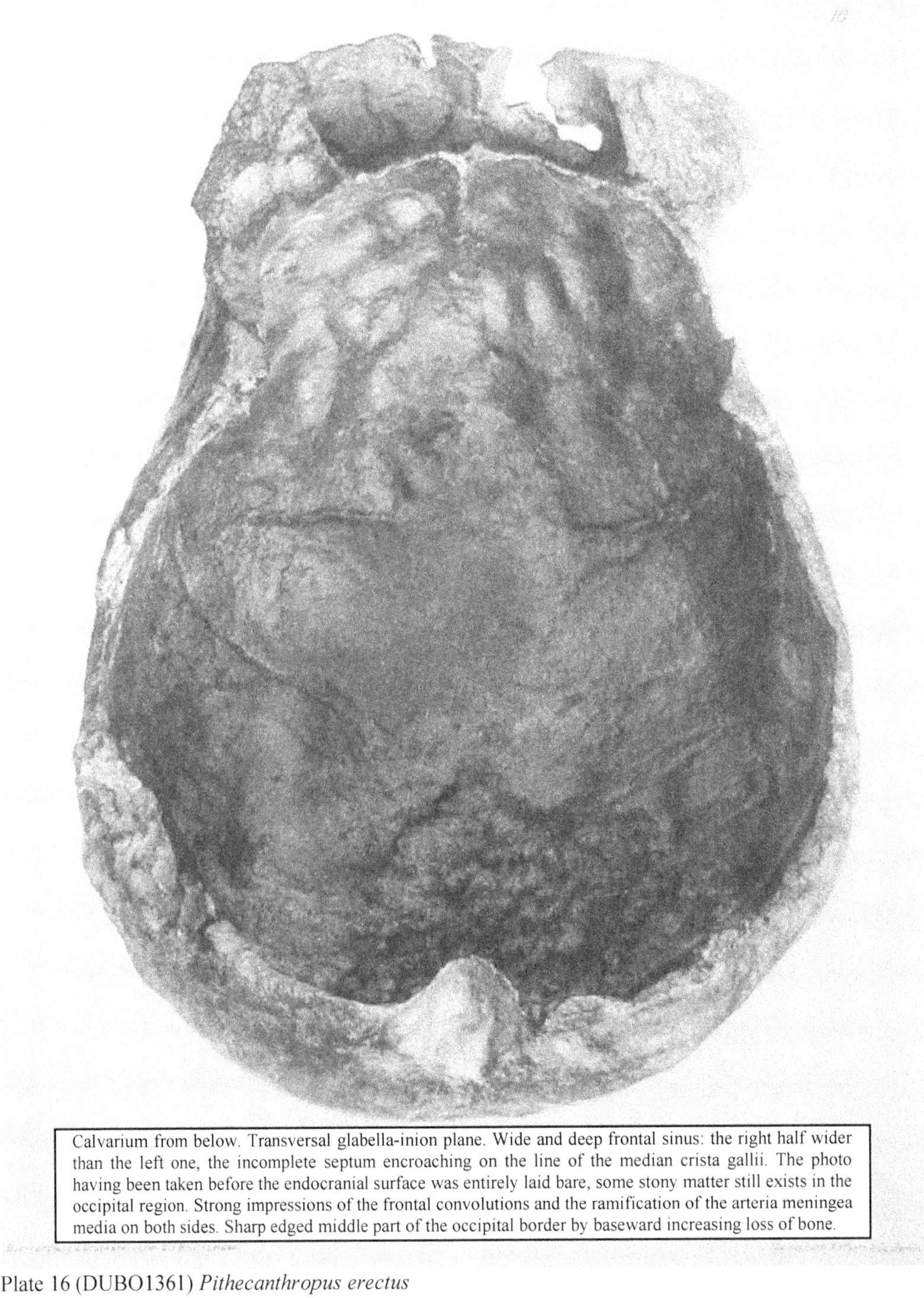

Calvarium from below. Transversal glabella-inion plane. Wide and deep frontal sinus: the right half wider than the left one, the incomplete septum encroaching on the line of the median crista gallii. The photo having been taken before the endocranial surface was entirely laid bare, some stony matter still exists in the occipital region. Strong impressions of the frontal convolutions and the ramification of the arteria meningea media on both sides. Sharp edged middle part of the occipital border by baseward increasing loss of bone.

Plate 16 (DUBO1361) *Pithecanthropus erectus*

Plate 17 (Top middle DUBO1408) *Pithecanthropus erectus*

Calvarium in front view. Perpendicular to the transversal glabella-inion plane. Glabellar region defective and corroded. On the right side, 28 mm from the middle, a supraorbital foramen. The swelling at the bregma exaggerated by the mentioned lost of the outer table of the parietal bones. Keel of the frontal bone. (DUBO1360)

Calvarium in occipital view. Perpendicular to the transversal glabella-inion plane. Inion evident, but still a little bony substance lost here. (DUBO1357)

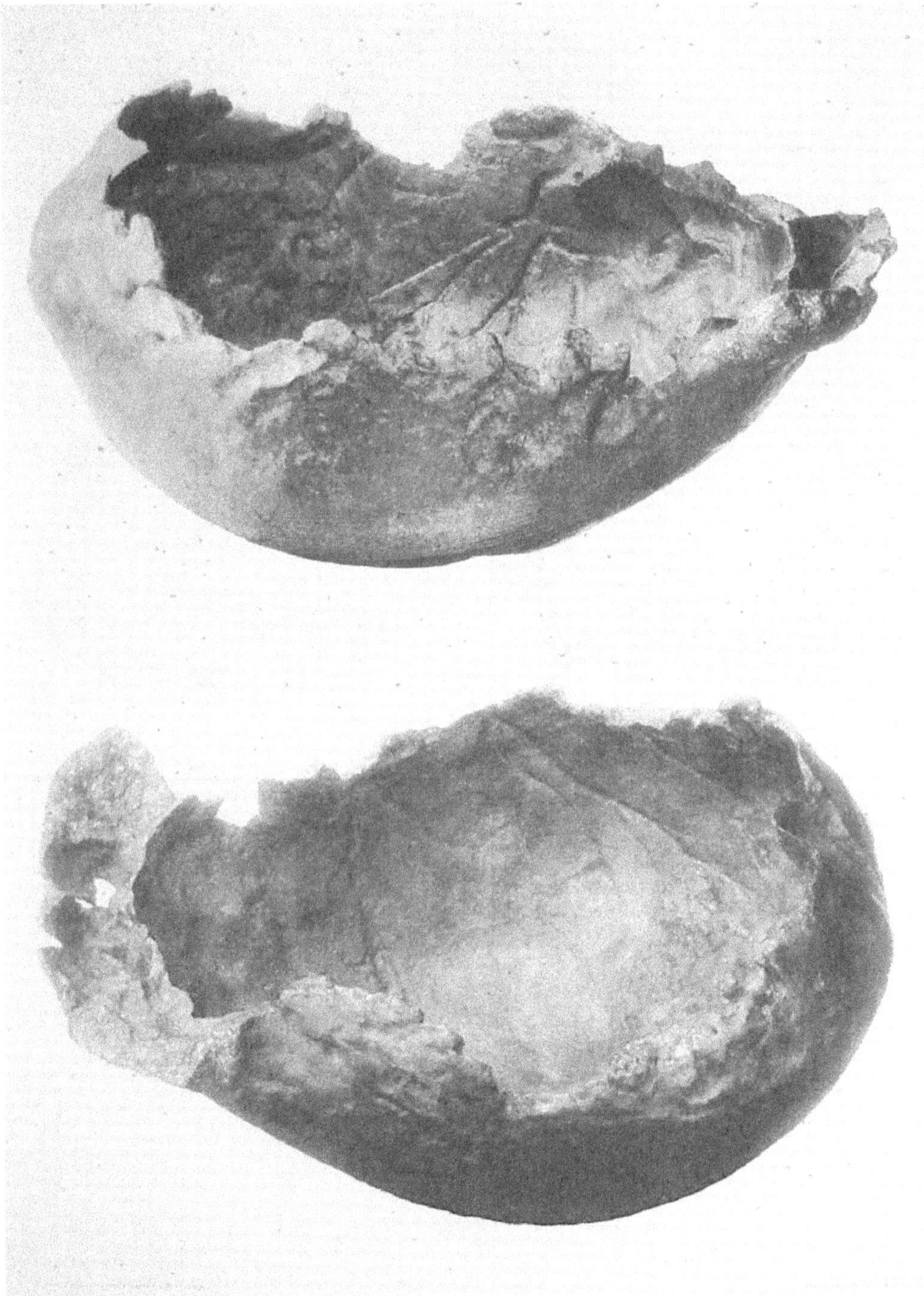

Plate 18 (Top DUBO1425, bottom DUBO1428) *Pithecanthropus erectus.* Tilted views from the side exposing the inside wall.

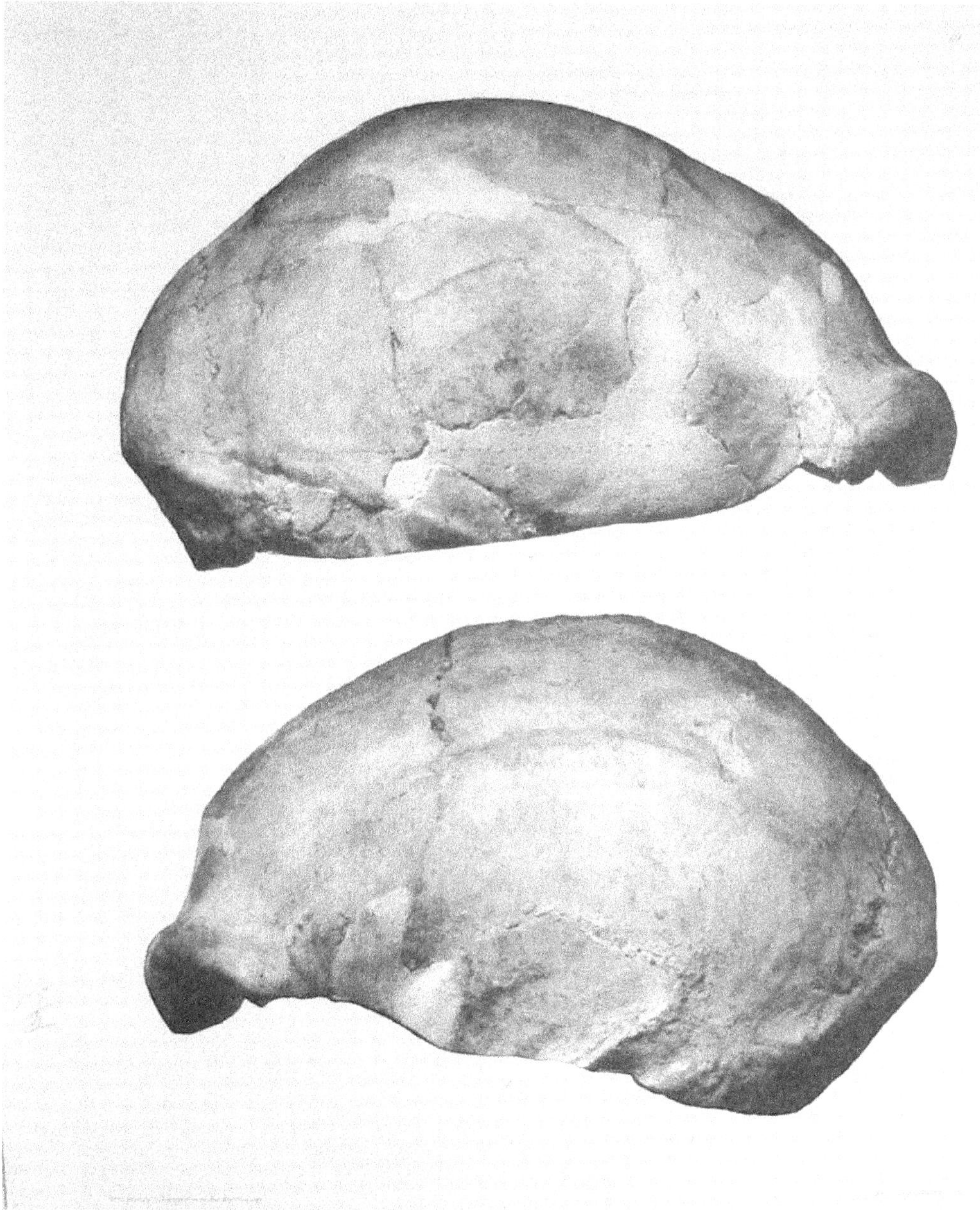

Plate 19 (Top DUBO1424, bottom DUBO1427) *Homo neanderthalensis* (Specimen from Spy)

Plate 20 (DUBO1423) *Homo neanderthalensis* (Specimen from Spy)

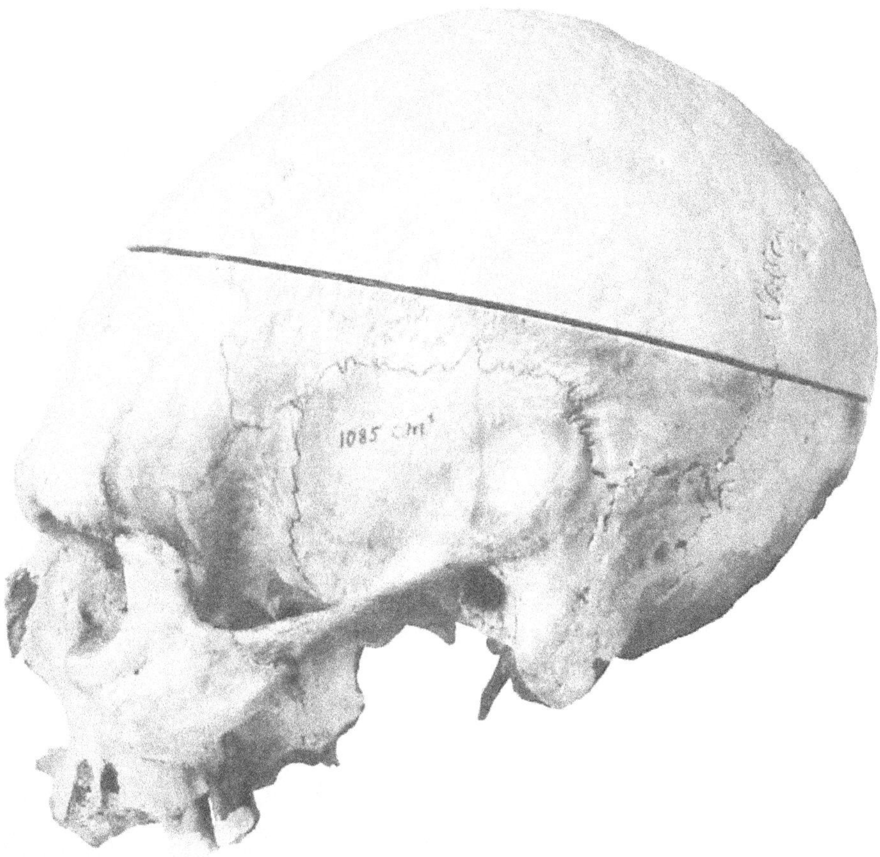


Plate 21 (DUBO1324) Microcephalic man (see also page 299)

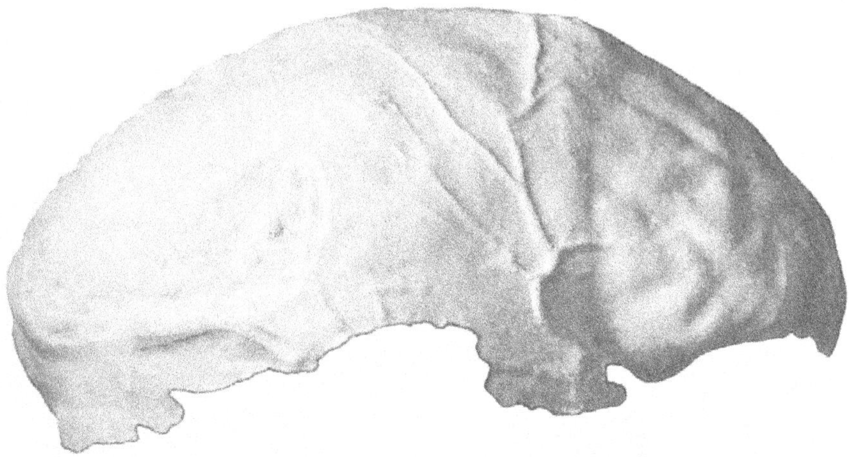

Endocranial cast. Right side. Most conspicuous the sulcus frontalis inferior, shaped like a '3' tilted to the left and which is prolonged in a fronto-marginal notch, and the encompassed ramus anterior fissurae sylvii. Strong system of sulcus frontalis medius and sulcus frontalis superior near to the frontal outline (2 to 3 mm distant). Clear sulcus intraparietalis (8 mm from the outline) and sulcus lunatus nearly vertical, above the internal asterion. The latter 2 mm below the upper border of the wide sinus transverses and 16 mm before the posterior limit of the sinus in this view. Fissura petro-squamosa clearly indicated.

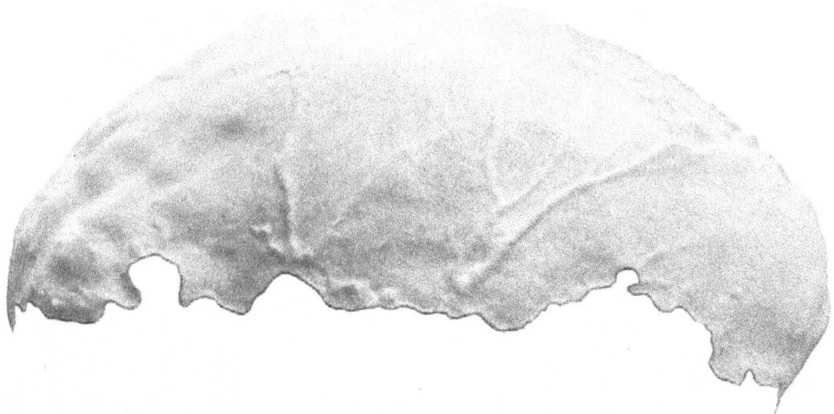

Endocranial cast. Left side. Ramification of the stronger ramus anterior and the less strong ramus posterior of the arteria meningea media. Sulcus frontalis superior 3 to 4 mm from the frontal outline. Sulcus frontalis medius, clearly homogogue with the sulcus rectus or principalis of the Apes, 8 to 10 mm, at least, from the outline. Sulcus frontalis inferior less distinct than on the right side, evidently somewhat nearer to the frontal outline, at nearest 18 mm distant in this view. Sulcus lunatus 20 mm from the occipital pole of the hemisphere.

Plate 22 (Top DUBO1318, bottom DUBO1323) *Pithecanthropus erectus*

Endocranial cast from above. Transversal plane of the cerebral fronto-occipital axes. Effect of trigonocephalism on the brain form. Exceedingly stron frontal convolutions. Evident intraparietal sulci. Sulcuc lunatus manifest, particularly on the right side. Sinus sagittalis and ramification of the arteria meningea media on both sides. Sulcus frontalis medius particularly developed as fronto-marginalis causing a notch, on both sides, in the frontal outline, about 20 mm from the middle line.

Plate 23 (DUBO1320) *Pithecanthropus erectus*

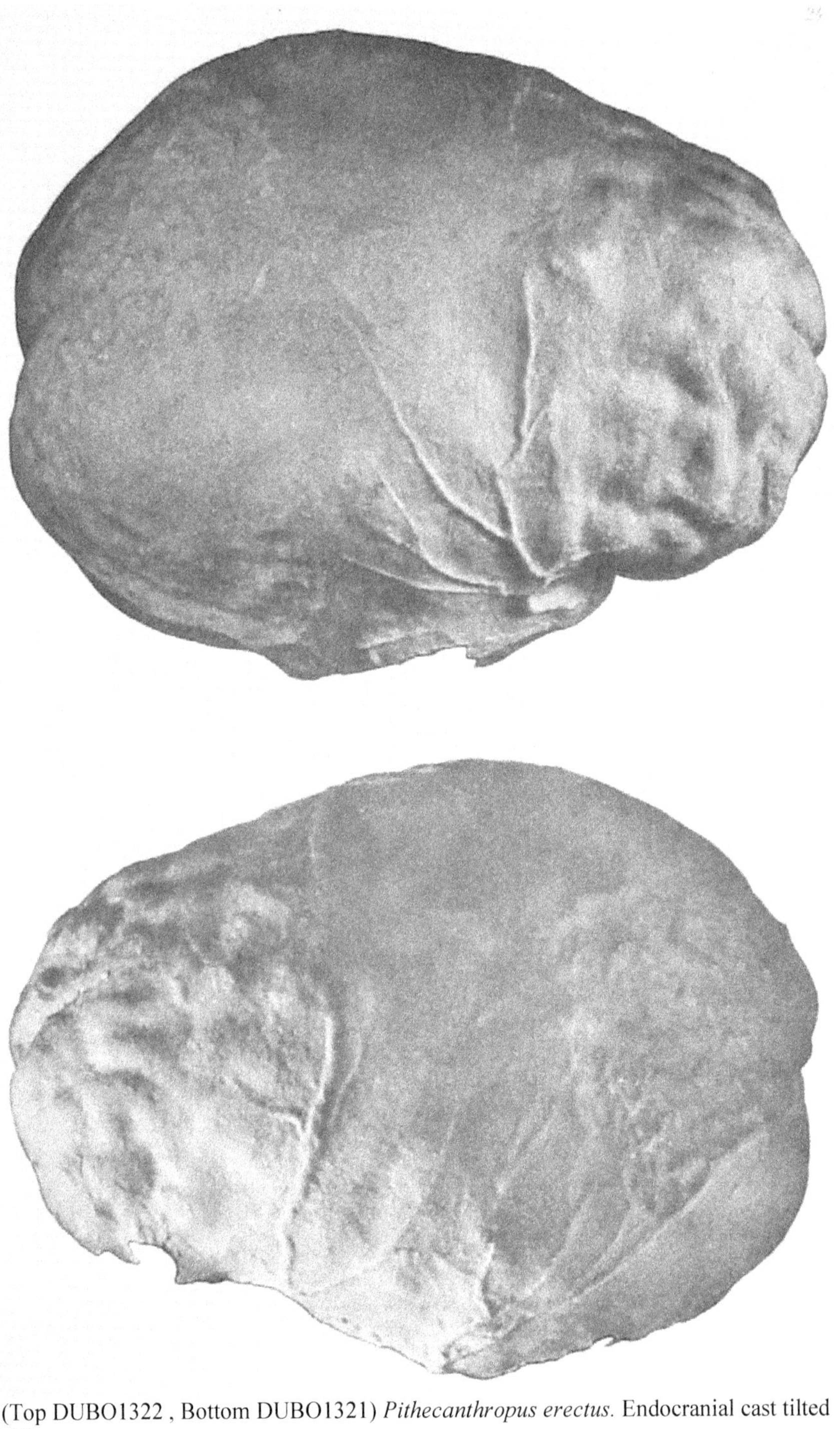

Plate 24 (Top DUBO1322 , Bottom DUBO1321) *Pithecanthropus erectus.* Endocranial cast tilted view from side/above.

Plate 24 concludes the plates that Brill had already printed for Dubois' book. Obviously the femur and the molars are missing and would have been depicted also at some point. Of the femurs a large fold out proof exists (below) which as a plate is present in his later publication on the looks of the femur.

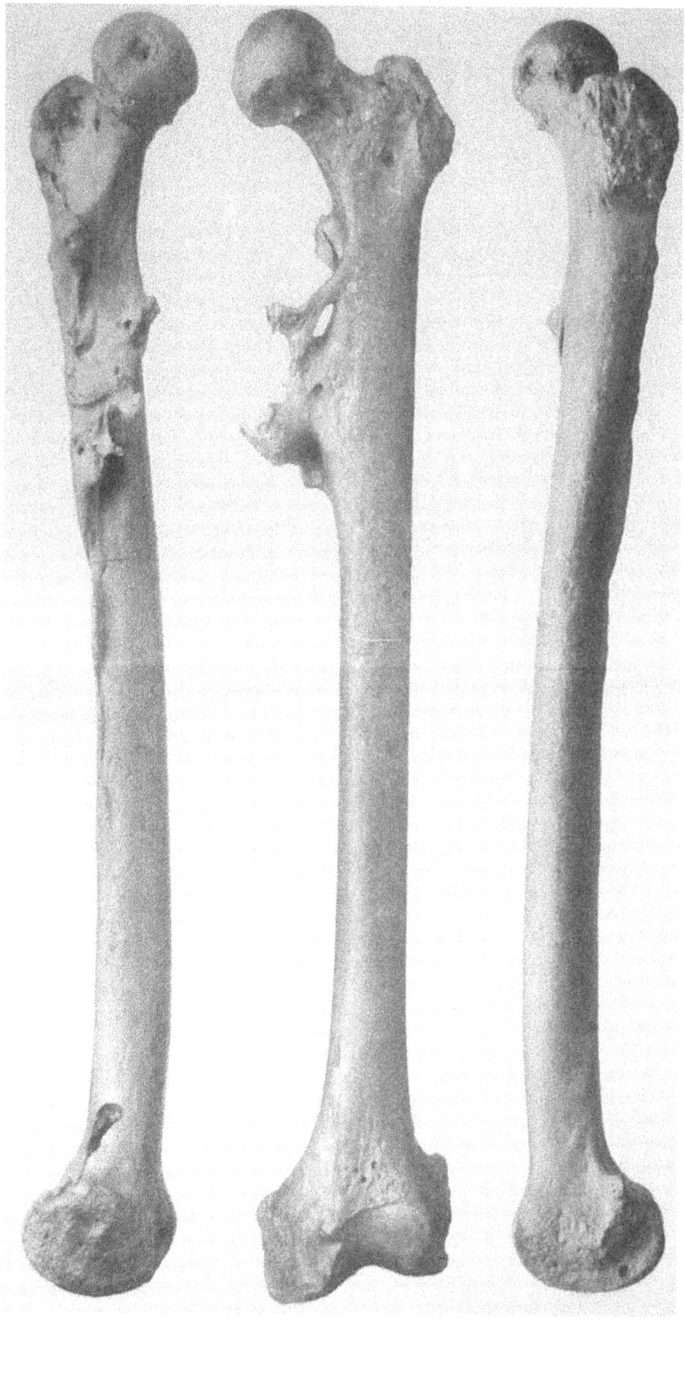

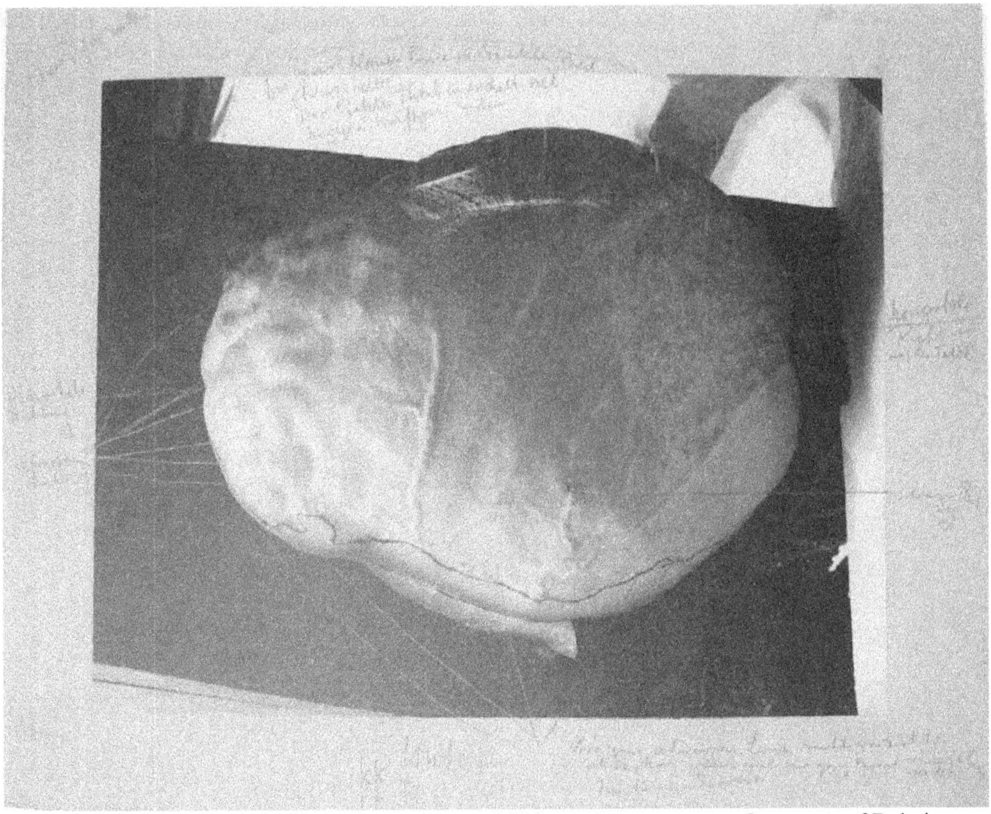

A proof of plate 22 of Dubois' book of an endocast of *Pithecanthropus erectus*. Comments of Dubois are in German (Brill had it printed in Berlin) and concern the direction this endocast is facing, parts that need to be slightly lighter, spots that need to be removed and the background which needs to be white: "*...Until the black line, do cover really careful...*"

*Character*

The reasons that Dubois gave to Brill in 1913 for not having finished his book were true enough but by no means complete. There were many more and a number can be attributed to Dubois' troublesome character. Dubois did not understand human behavior very well and as a result had great difficulty getting on with people. His empathic abilities were heavily skewed towards distrust and by treating people accordingly he not only saw enemies everywhere but created them 'en masse' along the way.

He was also meticulously precise and adamant in his objections to the smallest of flaws if he could find one. No doubt this was due to him as he not wanted to give his opponents any ammunition at all. With regard to his book many problems arose from this because he just could not agree with the proofs of prints and photographs as he could always find something wrong with it.

Although when they started the project in 1897 Dubois' standard was too high to be met by the technical possibilities of that time, these possibilities were developing rapidly, and De Stoppelaar, upon having seen the latest on a publishers congress in Germany writes to Dubois in 1898, that the firm Meisenbach, Riffarth & Co. would be able to meet his demands. Meisenbach is even prepared to send a photographer (*with* camera!) from Berlin to Leiden to photograph the objects provided they will be lying ready. The photographer is sure he will be able to make 10-20 photographs a day…

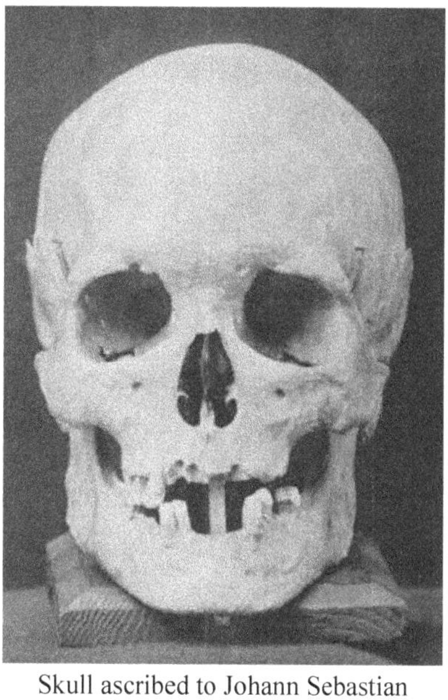

Skull ascribed to Johann Sebastian Bach, reproduction made by the firm Meisenbach, Riffarth & Co, Leipzig, shortly after 1894 (Dubois collection).

Clinching the deal were probably the reproductions of the prints of the skull of Johann Sebastian Bach that had shortly before been made by Meisenbach and which Dubois must have acquired via De Stoppelaar. In 1894, the church counsel of the Johannes-kirche in Leipzig had decided that the bones of JSB needed a more honourable burial than they had received after his death. They decided unearth them and although they were not very sure where to dig, an educated guess was made and the skull they decided on was JSB's was photographed before putting it in a tomb in the church. The reproductions of these photographs were made by Meisenbach and the very good quality of these prints advertised their standard of work which at that moment could not be approached in the Netherlands nor probably anywhere else in Europe or even the world; they had won many prizes for their work worldwide and were to win their largest prize a few years later at the world exhibition in Paris in 1900. De Stoppelaar therefore decides to have the photographs reproduced by Meisenbach and thus Dubois sends some of his material to Berlin, though hesitantly.

As was to be expected he is not at first pleased, Meisenbach is also not fulfilling all of his obligations, but being convinced they can't get anything better elsewhere he finally agrees in sending him all the material (with strict restrictions that negatives need to be returned) and by the end of 1901 finally prints are being made. Already four years further by then, De Stoppelaar must have given a sigh of relief for getting to that point, expecting to soon see the end of the tunnel. He did not realize he had only just left the harbor and there was a whole ocean of trouble to follow. Dubois may have consented but he was not pleased with the results. From then on he did not wholeheartedly cooperate anymore and the project was doomed…

De Stoppelaar feels the project slipping out of his fingers and seeks further support from his friend Karl Martin.

## *Martin*

Karl Martin on expedition to the Moluccas in 1892

Karl Martin got his PhD in 1874 at the University of Göttingen and came to Leiden to study the collection of Winand Staring, a geologist famous for his work on the geology of the Netherlands. In 1877 Martin was offered the chair of professor in Geology in Leiden and became in 1880 the first director of the newly created Geological Museum. The geology of the Dutch colonies was a primary focus of this museum and the main subject of his research. He was a very successful and highly esteemed member of society and could consider her Majesty Queen Wilhelmina amongst his private students. Martin could have become Dubois′ most powerful ally, but they did not get along from the start, or rather; Dubois quickly took a dislike to him after some regrettable incident and never could be persuaded to change his mind. “Reconciliation” just wasn’t a word in Dubois’ vocabulary.

The prize Dubois paid for his stubbornness was high. The fact that by the end of his career most of the scientific community had turned on him and considered him a maverick or worse has a lot to do with his paranoia-driven rejection of people like Martin.

Martin had supported Dubois from the go. When Dubois arrived in Indonesia he was an army doctor first. Martin was amongst the people whom Dubois specifically approached for support to persuade the politicians and army staff to allow him time and people to dig. Martin gave that support full heartedly but did expect one thing in return: he wanted the fossils that were to be collected to go to the new museum he had become director of, the Museum of Geology and Mineralogy. Many of the letter exchanges between the two revolve around that issue. Martin clearly states to Dubois that he is happy to support him even if he is not going to give the fossils to his museum, but makes clear that his appeals to the Government will have more force if he does them from his position as director of the museum where the fossils will end up rather than just from his position as professor in Geology stating the importance of paleontological research in general. Dubois understands and agrees to it. He promises that the fossils will go to the Geological Museum and frequently expresses his thanks to Martin for his diplomatic support.

Well known photograph of Dubois' veranda

Trying to find out what the incident could have been that led to Dubois' everlasting distrust has proven to be difficult. Rumour has it that Martin would have described a fossil of the Dubois collection without Dubois' consent but this is virtually impossible. In fact Dubois prevented that from happening by keeping all fossils with him in Indonesia in spite of repeated requests of Martin to send them to the Netherlands.

The exchanges reveal a dispute about ascribing species and 'the Siwalik fauna'. Martin had published that he had found a Siwalik fauna on Java, Dubois did not agree. He argued that the species he found were a fauna on their own, key species of the Siwalik fauna were lacking, others were here that were not reported in Siwalik. Martin argued that the certainty with which Dubois ascribed species names to fossils whilst in Indonesia without the possibility of comparison to specimens in the British Museum and the Museum of Calcutta was simply not possible. Dubois stated that some of Martin's determinations had been wrong and that he had ample material in support of his claims. They decide to stop fussing about this in their letters, both understanding that the other had valid points. It is therefore no surprise that Dubois' first move after Indonesia was to go and visit the Calcutta Museum and the Siwalik site. But this dispute was a scientific dispute and would (or should) on itself not have been a reason for personal distrust. Martin however kept requesting Dubois to send his material to the Netherlands instead of leaving it all on his doorstep (where it was quite literally).

DUBO0909: Last of the four 'veranda collection' photographs (DUBO0907 &DUBO0910, see page 15)

And this is probably where Dubois' paranoia kicked in; he could not imagine any other reason for Martin to ask that other than wanting to steal his data. Martin senses this distrust immediately for what it is and states as officially as possible that he will leave any crate unopened until Dubois will return. He has that letter copied and added to the museum archives with the specific notion to Dubois that he does so in order for Dubois to be able to fall back on that if need be. But the jinnee was out of the bottle and nothing Martin could do or say could put it back in. After Dubois had died Martin stated in an interview to Brongersma that he thought the distrust of Dubois had been caused because Dubois had heard that whilst in Indonesia Martin supposedly had said at the Royal Academy of Sciences that *Pithecanthropus* was an ape.

The first caretaker for the Dubois collection started off to become a complete lineage: sons following in the footsteps of their fathers of which this is one of the early "de Koning"s

When Dubois had returned to the Netherlands he had immediate problems. He had sent the fossils ahead to Leiden where they arrived one afternoon at Martin's doorstep, without anybody knowing they would arrive. Not that there was anybody other than Martin who would even wanted to have these fossils. The government was not that interested at all; they owned everything but were not eager to spend money on it. They even offered everything to Martin to work on. Martin made clear to the Government that surely Dubois should be allowed to work on the fossils that he had collected and subsequently Martin arranged a deal whereby he even declined part of the subsidy to his Museum in order to make that money available for the Dubois collection. In a rush a house was rented to store the (unopened) crates, but the amount was thus that they immediately had to look for some other place as the owner feared the floors would not support the enormous weight of these crates. When Dubois arrived in Leiden he started off with quarreling with Martin and accusing him of trying to steal his finds. Martin immediately sent for his correspondence, which he had carefully kept as he had foreseen this moment, and Dubois had to eat his words and admit that Martin had always supported him. He had taken on and housed the collection, a room was provided for Dubois in the museum where he could work on the collection and also out of the museum funds a caretaker was appointed to the Dubois collection.

Even after this incident Martin did not stop supporting Dubois: the money for the contract that Dubois had for his book, the 6350 guilders, was also deducted from Martins funds. As a consequence, though, Martin was in on all moves, he was always consulted on anything that had to do with Dubois or his collection. In 1895 Martin writes, again in official style, stating all he will do for the collection, it to be brought under a roof, etcetera, but with regard to the book he ends:

*.... I shall therefore propose to the Gentlemen Curators, to have the funding, that is exclusively intended for your report {the book on the Indonesian collection that is}, to be awarded to you personally, with particular regard to the fact that you have recently pointed out to me, in words that could not have been misunderstood, that you were not sure whether you could put trust in me.*

*I think that the agreement I propose now has a favourable chance of receiving your approval and hope you will acknowledge that this writing is objective and in your best interest.*

*Professor Director K. Martin*

Whatever the scientific dispute and whatever the personal differences, Karl Martin saw the importance of Dubois' work and the importance of the collection very clearly. He was in a position to seriously damage Dubois if he had wanted to, but kept supporting him for a long time in spite of the openly uttered distrust from Dubois' side. Furthermore Martin also read out some of Dubois' papers at the Royal Academy of Science, again something one would not do for his enemies.

The Dubois archives hold a letter of professor Lorié in which Martin is blackened, and which states that he (Lorié) would not put it past Martin to use dirty tricks to get at Dubois' fossils. There is also a curious slither of paper, which seems to be cut out of a letter, and which is unmistakably written in Karl Martins hand. It makes clear that Martin at some point also got frustrated with Dubois as he writes:

> *As far as the elephants are concerned, it might interest you to know, that my research on the stegodonts etcetera has been the immediate cause for the collection of Dubois (....) such as he stated to me himself BEFORE his departure to Indonesia. But Dubois would not have told you that, as he also will not have told you that the {money for the} report on the Dubois collection has only been awarded to him on my instigation - that was not something fitting in the mindset of this nervous man.*

We have no idea where this slither originates from. Martin outlived Dubois about a year and we know Brongersma interviewed him after Dubois died in particular to question Martin about these matters and may have thus acquired it and added it to the archives. But if someone has passed it on to Dubois during his life it will surely have aggravated the situation between them. We have no idea who, in that case, the culprit could have been, although Lorié is a likely suspect. When Dubois however turns to Lorié for support, as we will see later, Lorié burns Dubois down completely, hence Lorié did not end as the best of friends to Dubois either, (nor would he have been before, meddling like he did).

## *The death of De Stoppelaar*

De Stoppelaar had known Martin for over twenty years and Martin visited him regularly and the subject of Dubois book was always on the agenda. The first years Martin must have eased De Stoppelaar's mind by explaining to him how Dubois must have underestimated the task ahead of him. Martin had seen the sheer amount of material that Dubois had brought in and which could not possibly have been

Frans de Stoppelaar (1841-1906)
Director of Brill (1872-1906)
(Courtesy of Royal Brill, publishers)

dealt with in two or three years. But Martin also notices that far less is happening than what should have been done. The room he vacated in the museum for Dubois to work in is only rarely visited, so rarely that he asks Dubois to sit in the library during those few days he is there and he gives the room to somebody else. Martin even suspects Dubois of sending false bills for costs he claims to have made for his work on the collection but he can never prove it.

Meanwhile De Stoppelaar worries. A big project of his, an illustrated bible, has been a complete financial disaster. The company as a whole is not doing badly at all but the expensive failures along the way tend to fall on his side and Dubois' book is going to be a very expensive failure indeed if it is not going to be published. And all the time Dubois apparently is doing nothing. He always claims to be working, but at the same time he is on the front pages of the newspapers regularly with many other projects, distracting him from his book. And then there are the problems with Martin. This first became apparent to De Stoppelaar in 1900. The intention was that "the book" would be an issue in the series of the Geological Museum, the "Sammlungen", which was under editorship of Martin. Dubois however did not want Martin's name anywhere on the front cover, not even as editor-in-chief! Martin subsequently made very clear, if Dubois wants it in the series, he would have to put up with him as editor-in-chief, the cover will be as it always has been with his name on it, and he will get to see the last draft. If Dubois does not want my name on it, Martin stated, fine with me, no problem at all, but then it will not be in the Sammlungen. De Stoppelaar had become the go-between of these exchanges and had to put up with an exploding Dubois, who was absolutely furious when realizing that Martin would get to see his last version before publication. Another letter had to be written to explain that the task of the editor-in-chief in this case would not be to change anything to the actual contents but that he was only overseeing the continuity of the appearance of the journal. De Stoppelaar writes to Dubois that it does not matter to him whether he decides to put his work in the series or not, as publisher he is partial to it. But all this quarreling is not doing his health any good. Also people regularly come to his office to ask about Dubois' progress, not just Martin, but also people from the ministry, journalists, and clients that have subscribed in advance to this issue of the Sammlungen which has been advertized and is expected to be an important contribution to science. In 1903 he falls seriously ill and is absent for five weeks.

Dubois at about 40 years of age

On October 30th, 1903, De Stoppelaar sums up what Dubois has gathered in the six years previous:

*.....We have got now: 1st these 40 pp. 2nd ± 44 pp. which are currently in print 3rd a considerable collection of plates. - Should these be assembled to become a first issue than, of course, a title should be added - which should*

*be easy enough - but also, in my opinion, a preface, explaining your methods, which can be changed in a later stage, but for now will serve its purpose for the Ministry. - Finally, to underline your wonderful plates, a clarification that can be folded out, which could be printed in a smaller letter.*

*Maybe these propositions are {superfluous}, if so blame that on my ignorance. - The sole intention of this letter is to be armed and ready, when we face the Ministry. -*

So in 1903 there is still no more than a vague outline of what could be a first issue of a series of volumes. The same letter reveals that the order of the pages is not certain and both De Stoppelaar and Dubois understand there is in fact no book. However, to receive any funding from the Government something has to be done. Apparently Dubois is in fear of losing the funding altogether if he does not produce his first part before a deadline at the end of 1903. Brill on instigation of De Stoppelaar has spent thousands of guilders to get these plates printed, so both parties agree to make a dummy to send to the Ministry to fulfill their obligation to present a volume and receive part of the funding. But Dubois panics. He has his second set of 44 pages printed without page numbers (he does not know yet where the text will go). Brill agrees (everything to get the dummy ready in time) but they are not pleased, it means all pages have to go through the presses again at a later stage to add the page numbers. And then Dubois backs out. He probably does not trust the dummy to be accepted and renegotiates his deal with the government, and gets an extension, hence the funding remains available but the deadline is moved. De Stoppelaar despairs. He begs Dubois to send him the "missing link" of his monograph. Again he has been led by the nose into printing material that may never be used. He finally must have realized that Dubois just cannot be trusted and starts to change his attitude. Trying to be more businesslike he refers to his duties as director, his responsibilities towards the company and tries to put pressure on Dubois by explaining that questions are being raised by members of the executive board. Dubois simply ignores him and does not even bother to answer his letters.

Prof. J. Lorié (Courtesy J. Reumer)

After a year of being ignored De Stoppelaar raises the stakes and sends Dubois notice that he will no longer wait, he will simply take all the plates and the text which are there, bind them in a cover and publish it. On the cover he is to write a short note stating, that without it having been the responsibility of the company, this publication has been delayed, and that he hopes soon to be enabled by the author to publish a sequel. Should Dubois wish to comment or should he have another proposition he will have to reply within 8 days. De Stoppelaar is even willing to go to him should he wish to discuss this man to man. But this year, 1904, something has to be published…

Dubois seems to be cornered. He turns to his friend Lorié for support but gets an unexpectedly blunt reply. Lorié dishes him all the dirt that is going around

about Dubois in particular that he is stalling publication in order to keep receiving paychecks from the Ministry[4]. On top of that Lorié writes that he agrees with the claims of Brill, he understands that the minister of Colonies does not pay up, and that Dubois is only getting what he has been soliciting for. Surely Dubois should understand that if he gets into the newspapers with all his other activities Brill is not going to be pleased?
*....What is it with those metamorphoses of you anyway, which just keep coming? 1- Medic, 2- Comparative anatomist, 3- Paleontologist, 4- Geologist, 5- Hydrologist, 6- Civil engineer. Next year you'll might as well be heading a machine factory, who knows?...*

It is a blow below the belt to which Dubois writes Lorié a furious response, but that does not help him much with his immediate problems with his publisher and on that front he concedes to a meeting.

The meeting is man to man but De Stoppelaar afterwards sums up a kind of historical overview of the events leading to the meeting and its outcome in an emotional letter. How the initial financial problems to publish such a work were lifted by the support of the Government and Prof. Martin who gave up on part of his funding for it. How people kept coming to his office to inquire after the progress on it, and though in the first years he could answer that the plates were being made or text was being printed, the book never appeared and Prof. Dubois was always otherwise occupied, of which the newspapers gave ample account. How this al saddened him and had cost him a lot of money *...for thou art also mortal, and who will finish it after you're gone?...*

And how Dubois stopped answering his letters and he decided to write that he was going to publish, just to provoke a response, which he was pleased to see Dubois did. De Stoppelaar ends:
*....everything can still turn out well, if you stick to what you have said, that you see this book as a part of your life's work. Truly, not a triviality! That people are gossiping about it, you know that as well as I do, but let us proof by the deed that the book will be there, as you have promised now, and all voices will be silent. "Le mieux c'est l'ennemi du bien". Give what you have got; it is such **valuable** material, which you have collected at the expense of large sacrifices, do let it be published under your name, and in your words.*

But whereas De Stoppelaar had given in to Dubois good intentions, the ministry was not so easy to win over. De Stoppelaar seems to think he is in a winning mood and sets himself to write to Dubois to overcome what he thinks is the final hurdle that needs to be taken to get to the end line: reconciliation between Dubois and Martin. Martin must have been in on it and willing to support the publication of the book anyway he could, but, he also wanted his prize: the Dubois collection in his museum. De Stoppelaar writes Dubois how he has been working with Martin for 30 years, how he always has been honest and loyal and of good will, whatever other people might say. He proposes a conference with the

---

[4] Dubois was officially still in service as officer of the army when in November 1904 he is being ordered back to Java. Thereupon he requests to be discharged on medical grounds which results in his honourable discharge in February 1905.

three of them, where all differences should be openly discussed and a plan to rescue the funding should be made.

But his request is in vain. De Stoppelaar completely underestimates the loss of face that would have meant to Dubois and does not see that he is only feeding Dubois' paranoid suspicions about his own loyalty towards Dubois. There are two more letters from De Stoppelaar's hand from which it becomes clear that for 1904 Dubois again negotiated an extension for the subsidy. November the 9th Frans de Stoppelaar writes his last known letter to Dubois in which he says that he will patiently wait and has new hope but the archives of Brill are void to Dubois' name for the next one and a half years. De Stoppelaar regularly falls ill again and by June 1906 his disease takes a turn for the worst. His fellow director Peltenburg writes to people like the famous Dutch politician Domela Nieuwenhuis and the eminent arabist Snouck Hurgronje about the state of his dear friend, that he has been brought to Amsterdam to undergo surgery of the stomach by Professor Rotgans. The surgery goes well, there is a brief moment of hope, but two days later, the 8th June 1906, De Stoppelaar dies at the age of 65. People around Dubois apparently need to have a strong stomach, as father Bernsen would find out to his regret almost exactly 25 years later.

# Distractions

In this chapter we lead you away from the storyline of Dubois' book, just as Dubois was continuously led away by what we now tentatively call 'distractions'. Pat Shipman used the word 'diversions' but we feel distraction is more to the point (surely arguably) as the many other occupations took him off track from writing his magnum opus. But some of these distractions were very important to him and would not have been seen as distractions by himself at all, but rather as main topics.

On arriving back in the Netherlands he was soon in need of money and accepted in 1899 a professorship in Geology at the University of Amsterdam. This came as a surprise to some, as it was obvious that this would not facilitate the work on his collection. But the obvious, a professorship in Leiden, was probably not feasible, as there was already a professor of Geology in Leiden, Martin, who was not about to leave anytime soon. Hence Dubois had little choice but to accept and settled himself in Haarlem which was well connected both with Amsterdam and Leiden. The job brought new things on his path, students, teaching and new insights in many directions...

DUBO1280 This 'cage' or stereorthoscope, as Dubois calls it, is specially constructed by himself and photographed to judge the extent to which the distance towards the lens and the quality of the lens itself influenced the measures that could be taken from the photograph. He used it to align the subjects he photographs within a particular viewing angle by placing the cage over the subject while focusing the camera after which he removed it while the negative plates were being exposed.

## *Photography*

We stated earlier that Dubois' standard for the quality of his reproductions was higher than the technical possibilities in early 20$^{th}$ century would allow. This does not do right to the marvellous lithographs that could also have been made and were common in scientific publications in the centuries before. They were sublime and very accurate but around 1900 they were just outdated. Any serious scientist would have to use the best method available: photography. Photography was considered a representation of the truth and therefore more suitable for science than an artist's representation of that truth. However, as Dubois very soon realized, there was very little 'truth' in photographs at all. The position, the lighting and the quality of the lenses did not automatically construct a true image, but far more often a distorted one. And he could not have that, the measures on his photographs needed to be absolutely spot on.

Nobody in his right mind will after he has taken a picture of his child check whether the distance between the eyes matches relatively to the distance between nose and chin. But Dubois must have done exactly that with the pictures of his fossils, and his measurements in all dimensions had to fit.

A further complication of that time was that photographic enlargers were not yet being used to make prints. Prints were made by putting the negative directly onto the photographic paper and then lighted, to create a so called contact print. Photographic enlargers had already been invented but they were merely used to make larger glass negatives from smaller ones. Without an enlarger it is particularly difficult to print

DUBO0454 (left) and DUBO0456 (right) Photographic experiments in progress: to photograph a skull, like from this horse, without any distortion due to perspective, quite some distance is needed (more than would be readily available indoors it seems, although the lighting could also have been a reason to go try outdoors. However with a lens of limited capacity like used here (probably his 64 mm lens but certainly one with less than 100mm focal distance) the object itself becomes too tiny to be useful.

something exactly at life-size because then you need to get the image at exactly the right size on your negative first. The same difficulty also holds if you want half-size or any set size in advance.

On many photographs it is very obvious that Dubois is dealing with these issues. First of all distance is needed. Distance is the only variable to influence the distortion you would get because of simple perspective. You need to be at least 2 metres from an object to let a centimetre in the middle of an object appear equally large to a centimeter at the edge of that object (see fig. 1)

DUBO0956 Experimenting in progress. A mirror and a white canvas have been set up to reflect more light to the subject from the left. But notice how the wall is visible through the mirror: apparently the mirror was added whilst the photograph was made or in between two separate bouts of lighting the negative.

On some pictures he mentions the use of a 64mm Steinheil Antiplanetic Group Lens which in those days would have cost him \$95,- to buy as we know from an advert from that time. It was advertised as especially good in avoiding distortions of any kind (though in fact it can produce a coma: a kind of hazy ring around at the edges). We know however, that Dubois also used a lens with a focal length of 1500 mm and a distance to the object of three metres. If you want a 1:1 image the maths of optics teach us that Dubois must have concluded he needed a camera that also was 3 metres long from lens to the negative. He built such a camera himself for the purpose, but alas it was destroyed after he died, so the exact features of this apparatus will remain unknown. We do not know for instance where he would have got the lens from that he needed. A good 1500 mm lens would have set him back substantially more than \$95,- if he would have had to buy it. But we just have the pictures to show that he was experimenting trying to solve the problem to get pictures in 'true' size.

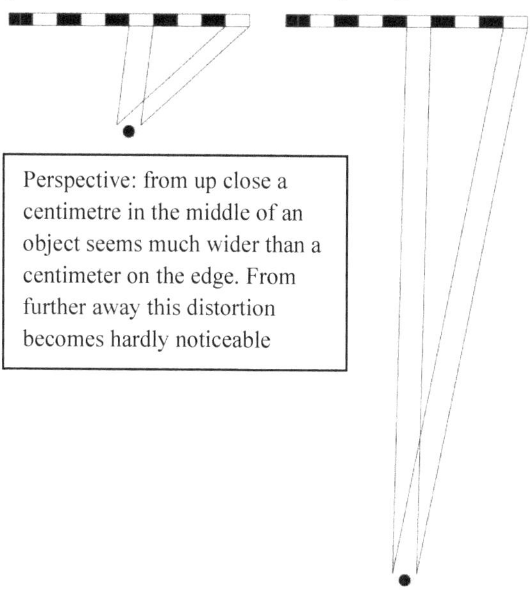


*Figure 1.*

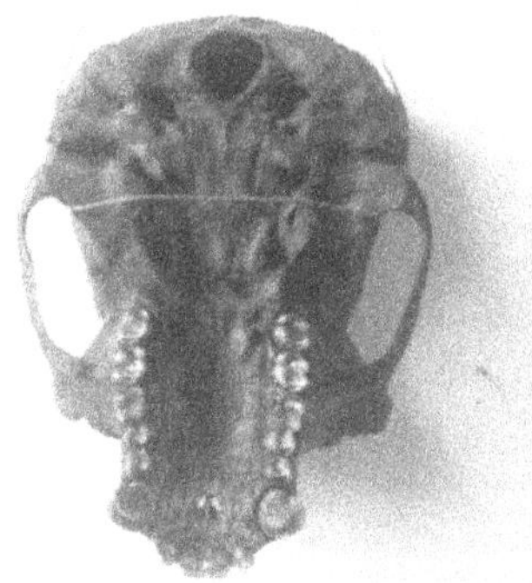

DUBO0302 (left) This gibbon skull *(Hylobatis agilis)* is hanging by a thread to accommodate the huge camera Dubois had build and with which it was not possible to photograph an object from above: there was simply not enough room between the floor and the ceiling.

DUBO0762 (bottom left) Dubois even dared hang his precious by a thread…

DUBO0882 (bottom right) The thread by which the *Pithecanthropus* femur is hanging has been (partly) scratched away from the negative.

We have learned that recreating these photographs nowadays will not meet approval of the staff currently responsible for these fossils…

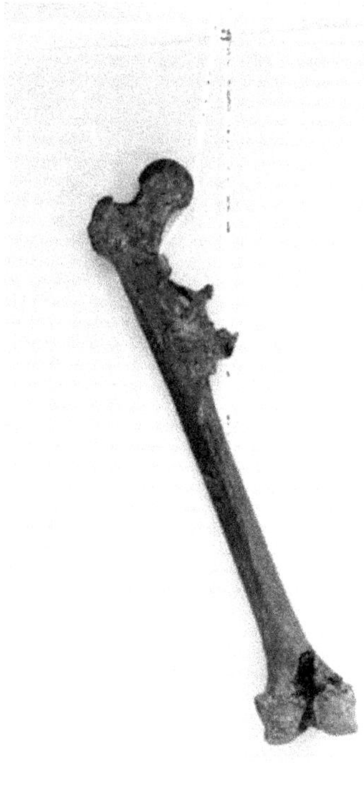

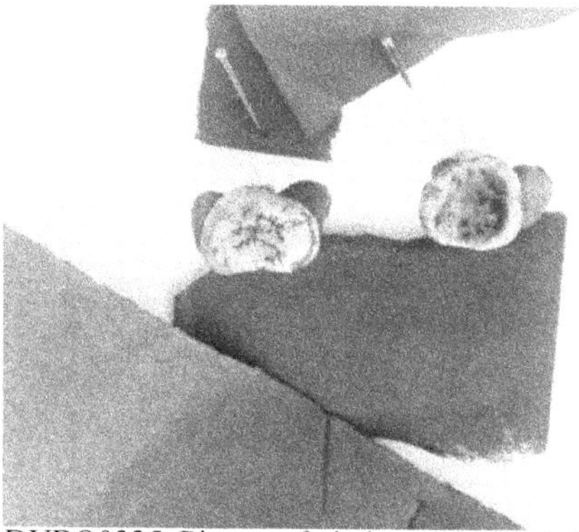

DUBO0325 Pieces of dark paper are being used to contrast the white enamel of Dubois' *Pithecanthropus* premolars. It did not bother Dubois that the picture as a whole was not very exciting as it would only have been made for a publication. The publisher would turn everything irrelevant to white.

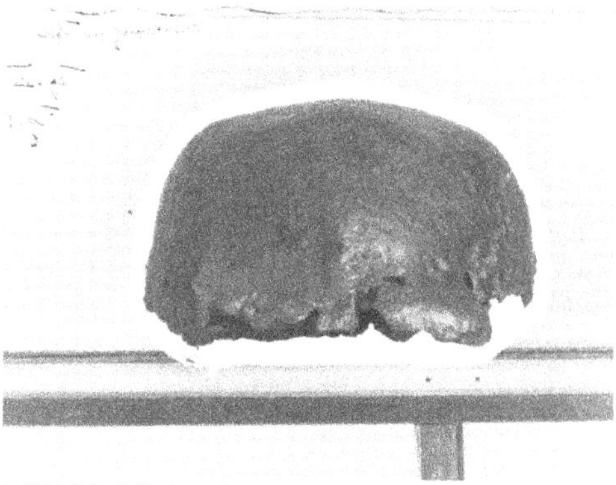

DUBO0728 On the negative of this picture the skull cap of *Pithecanthropus* is carefully bordered with red paint. In photographic darkrooms red light was being used which did not influence the photosensitive layers. This way Dubois ensured that the outline of the picture appeared in print exactly as he wanted it to.

DUBO0788 Aluminium foil is being used to create a diffuse light on the left side of this braincast

DUBO0706 Photographing through a tube, possibly to create a slide to be used in a 'magic lantern'.

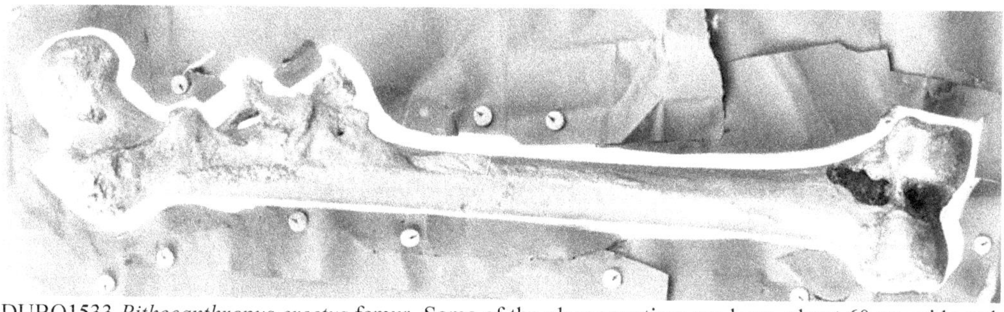

DUBO1533 *Pithecanthropus erectus* femur. Some of the glass negatives are huge, about 60 cm wide and about 40 cm high, to accommodate lifesize contact prints on photographic paper. These pictures were intended to be fold outs in the book that was to be. Surrounding the subject with dark paper prevents the white haze that otherwise would fall over a relatively small subject against a large white background.

*Fool's errands*

DUBO1437 the plaster cast that had been send to the Netherlands

Being famous, as Dubois certainly was in his days, you are being expected to respond to anything remotely in your area of expertise. The first big thing however started off very promising. News came from Indonesia that another *Pithecanthropus* skull had been found on the Trinil site. The finder, Heberlein, immediately went to the local press announcing the big news. The government was notified of this find of great importance. Press and Heberlein himself contacted immediately the Netherlands to receive Dubois' opinion on the subject, and in no time all newspapers in Indonesia and the Netherlands were buzzing about the new find. It took a bit of time before Dubois got to see what he had actually been commenting on, and it proved not to be a *Pithecanthropus* at all, it was not even a skull, but instead part of the cap of a joint in the front leg of a fossil elephant. Old yes, interesting no, in fact the finder probably got it from a rubbish dump of the original site where it had been left behind as not even worthy to ship back.

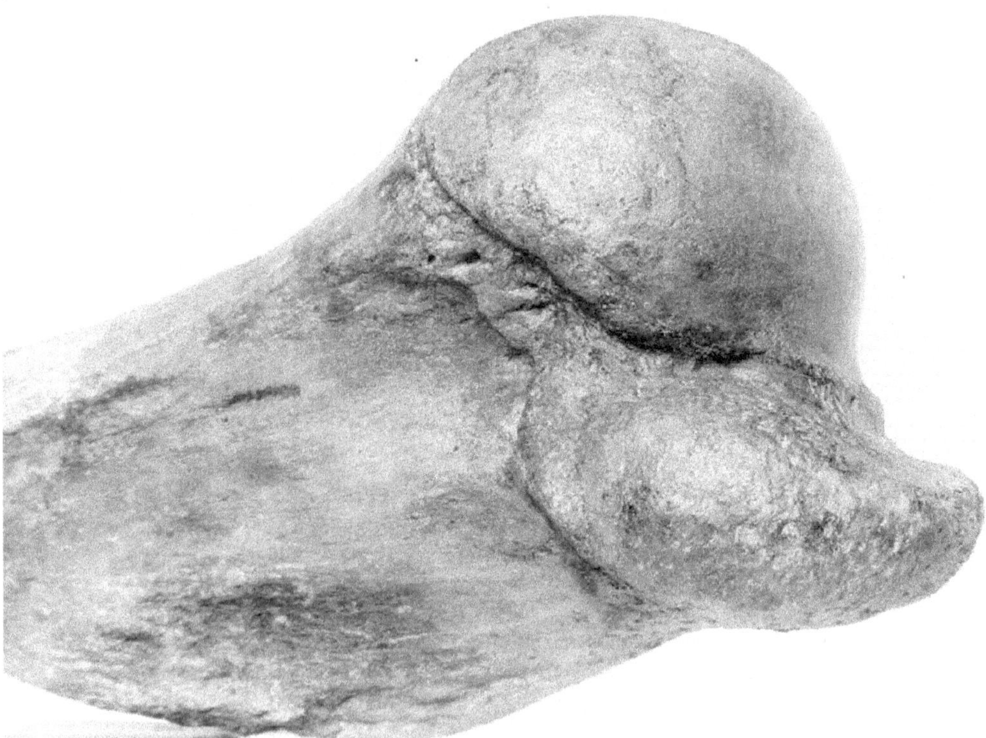

DUBO0674 what this particular bone looks like when complete

Dubois felt uncomfortable with the whole situation, though he was not at fault, but still sought reassurance of his scientific friends like Max Weber, to express that he really could not help having been part of this hype.

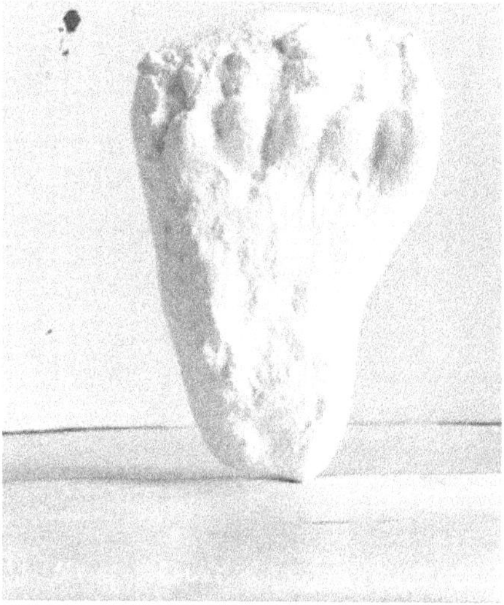

DUBO4232 cast of footprint of orang pendek

But on the other hand Dubois was a willing target for journalists who wanted a comment on some find from here or there. He even commented on the footprint of the legendary 'orang pendek' (short man), a 'yeti'-like creature that is supposed to roam the forests of Kalimantan, and that, as Snouck Hurgronje in those days pointedly put it, "is always one day length of travel further into the forest than the traveler is willing to go....". A cast of the footprint was sent to the Netherlands and ended up on Dubois' desk. He was however more careful with his statements this time, and described it as probably being of a Malayan bear. In the light of the recent finds of the small erectoid *Homo floresensis* it is noteworthy to realize that for more than a hundred years in Indonesia their mystery beings were not large impressive beast like yeti's, Sasquatchwans and Loch Ness monsters, but small men or men talking rubbish like orang pendek and orang letjo and it is no wonder these beings have been brought back into the limelight. We can certainly help that cause by presenting here the photograph that Dubois had made of the orang pendek footprint.

We are pleased however to not have to report any comments of Dubois to the following pictures. But they were sent to him nevertheless: all paleoanthropologists probably attract weird skulls.

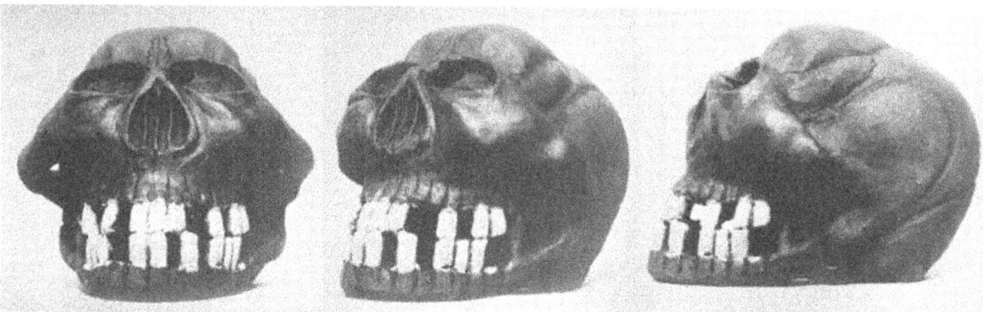

Closer to home matters were no different. The archives hold several letters reporting finds of skulls in clay pits along the river Meuse, in particular in the Tegelen vicinity, but never anything thus old to be interesting.

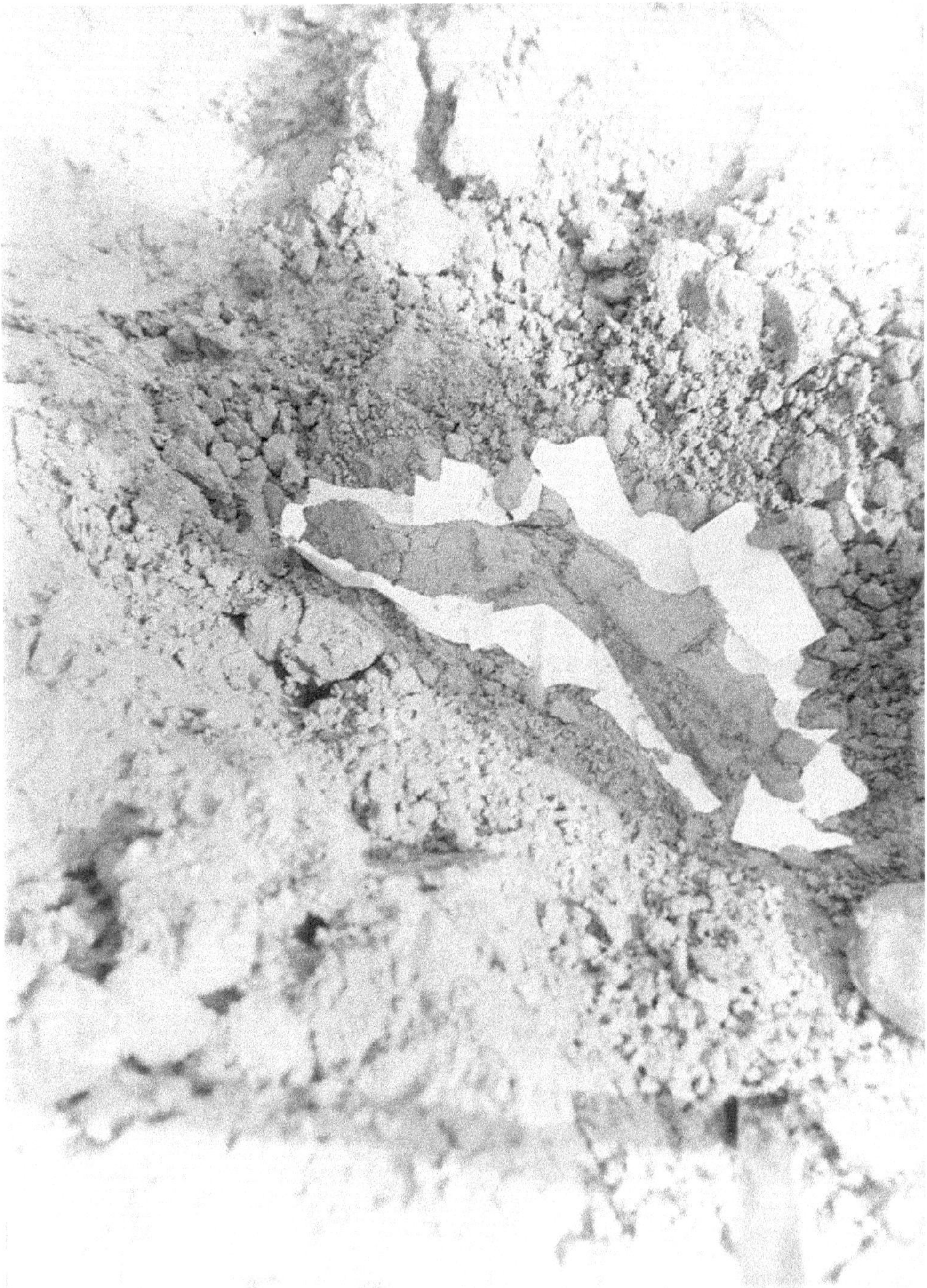

DUBO1039 Though it may not be obvious on first sight, this picture is a very spectaculair find: surrounded by pieces of white paper you are looking at an *in situ* photograph of a partial antler of the famous giant deer, *Eucladoceros tegulensis,* of Tegelen.

## *Tegelen*

DUBO0747 Quarry Canoy-Herfkens 1904

Eugène Dubois originated from the province of Limburg, the utmost southeast part of the Netherlands, in many respects different from the rest of the Netherlands and by no means part of 'Holland'. And although he had been to the other side of the world and his work remained mostly on the other side of the country, his heart must have stayed in Limburg, where he bought a large piece of land, built himself a second home and spent as much time there as he could. In that era large pieces of hitherto barren land, mostly heather, were being brought into cultivation and the need for building materials made people explore for clay and gravel along the Meuse, the river which 'defines' Limburg. The area had been renowned for its quality clay since Roman times (the names Tegelen originates from the Roman *tiglia:* roof tile) and had at that time a booming ceramics industry.

August Canoy (Courtesy of Mrs C. Canoy-Dankelmans, his daughter-in-law)

And lo and behold, in these clay quarries, fossils were being found, in Dubois' back garden, so to speak, and as he was by now the obvious specialist to be consulted on fossil matters, he acquainted himself with these sites, only to find that they were in some aspects not so different from the situation he had encountered in Indonesia. In fact, on occasion of the anniversary of Dubois' hundredth birthday, Ralph von Koenigswald stated in an interview that what Dubois really was after in Tegelen was to find his Dutch *Pithecanthropus*, and although, to our knowledge, Dubois has never said it as such, this seems to be a fair assumption (Folia Civitatis 11:16, 24-1-1958).

For a long time it has been assumed that this all started off with the brother in law of Dubois' brother Victor, August Canoy. But there is an error in this that has been repeated in publications since 1941. August Canoy was NOT the brother in law of Dubois' brother Victor. Victor was married to Maria Carolina Barbara Canoy and August had a sister Maria Bernardine Adolphine Hubertine Canoy and both were called Marie but they differed 18 years in age and there family relation was not that close: the great-grandfather of Victors wife was the great-great-grandfather of August, hence Dubois' relation to August was in the sixth grade on the in-law side.

DUBO1394 *Eucladoceros tegulensis*

DUBO1395 *Cervus rhenanus*

However they lived thus close to each other and both were from the local respectable upper class that they surely must have met occasionally. August Canoy owned one of the quarries in the Tegelen area, the quarry Canoy-Herfkens. He noted that bones were being found and was aware that a nice collection had been assembled by a student of medicine Laurens Steijns. Dubois was already aware of finds in Tegelen in 1897 but is still busy with his Indonesian fossils and only visits the sites first with two students in 1902. Soon pressure is being put on Steijns to part with his collection for science' sake, but Steijns does not yield immediately. He sees himself acquiring an academic position because of his collection, much like Dubois had done with his Indonesian collection. Dubois has to set him straight and explains that chances for an amateur medicine student to get a position at the university are not realistic as there is hardly any money to hire staff at all and as there are already fully trained paleontologists with successful digs on their names who are unemployed. Via Canoy also local pressure is rallied and it ends up with the Steijns collection being donated to the Teylers Museum in Haarlem, the oldest museum of the Netherlands with a world famous collection of fossils amongst which are best known the *Homo diluvii testis* and an *Archaeopteryx*.

The contacts that Dubois built up with the local quarry owners amount in a large collection of fossils from Tegelen that all end up in the Teylers Museum where Dubois had become curator. In 1904 Dubois publishes a first paper on this matter: On an equivalent of the Cromer Forest-Bed in the Netherlands, followed the next year by a description of *Eucladoceros tegulensis*, the large deer of Tegelen and *Cervus rhenanus*, the small deer of Tegelen. Pieter Tesch, later head of the National Geological

DUBO1023 Plant seeds from Tegelen

Department, did not agree with Dubois' dating of the strata and prompted Clement Reid and his wife Margaret to contact Dubois to be able to examine the plant material. By 1906 Dubois arranges a borehole to be made at the Canoy-Herfkens quarry and the results lead him to conclude that there had to have been a prediluvial ice age. This was a remarkable conclusion at that moment as until then the Pleistocene had been defined as being the ice age and Dubois now told that there had to have been another ice age *before* the Pleistocene. On 'invitation' of Dubois Reid studies the plant seeds that had been collected from the clay. These finds lead to the conclusion that the layers in Tegelen were not compatible to the Cromer-forest bed as Dubois had published, but were something entirely different altogether, which is subsequently being named by Reid and Reid in 1914 as the Teglian (= Tiglian).

DUBO0749 Stratigraphy at quarry Canoy-Herfkens

Tegelen fossils and the dating of the Tiglian as such are nowadays still being debated. Dubois was certainly not correct in the layers he all assigned to be Tiglian as was already muttered in his day by people who had worked in the field and had seen the differences between the types of clay in the different quarries that according to Dubois were all yielding Tiglian material. Later pollen research however initially more or less confirmed Dubois' hypothesis. This stance was regretfully defended by the pollen researchers with a zeal that can hardly be considered scientific and at the cost of anyone skeptic to their views. Only recently the two groups are coming a bit closer to each other as the body of evidence against the pollen confirmation has become thus large that the zealot side is starting to give in. Dubois has been at the root of this controversy as for many Tegelen fossils it was never entirely sure where they originated from. There were six odd quarries in the vicinity of Tegelen, but none of the original sites exposes outcrops anymore, and none of them were examined at the time in such a way that the disputes of today could be settled. In fact, hardly anything of the actual stratigraphy of that time had been recorded in a way useful to current research.

DUBO0751 Quarry Maalbeek east of Belfeld

Had collecting been done in a more organized way many problems would have been avoided. But as it was the local workmen just handed over their finds to the quarry owners and once every now and then they were picked up and/or sent to Haarlem. The Dubois collection turns out to hold a number of photographs of the original stratigraphy of the type locality of the Tiglian that were hitherto unknown and might spark of some new insights for future research. The archives also turned out to be holding a number of photographs that Dubois did not make himself but acquired as they were probably being sold as souvenir snapshots (see appendix III, page 182). Amongst these there are also some on which the original stratigraphies of several quarries have been depicted. Aside to Tegelen Dubois also took an interest in other locations in the Netherlands where geological phenomena could be observed that where linked to the ice age, like in Maarn and 'de Hondsrug' in the province of Drenthe.

DUBO0744 (16 May 1904) Dubois: "Upright and wrinkly layer of loam in the sandpit in Maarn, the loam is beautifully layered, 20 cm wide at the top, 80 cm below near the rails…"

DUBO1084 (9 December 1904) 'de Hondsrug'

As was the case with the Indonesian collection, Dubois did not have the patience to make extensive descriptions of all the species in the fauna of Tegelen. These descriptions were being done by his assistants in Leiden and Amsterdam: Father Sanctes Bernsen wrote his dissertation on rhinoceros finds in Tegelen and Antje Schreuder wrote her dissertation on beavers from that location. Bernsen continued the description of the mammals in a series of publications until his untimely death in 1932, after which the work was finished by Schreuder. Antje Schreuder became highly esteemed worldwide for her knowledge on Pleistocene mammals.

Antje Schreuder (collection Geological Museum Artis)

Sanctes Bernsen surrounded by his fellow friars in Heerlen who awaited him to celebrate his promotion after a cum laude defense of his dissertation on the Rhinoceros from Tegelen.
Courtesy of KDC/KLIB Nijmegen.

## *Saving the Netherlands from thirst*

With the industrial revolution well on his way and the mechanization of the agricultural work driving people from rural areas to the towns, the cities in the West of the Netherlands were expanding rapidly. The population growth around Amsterdam brought with it a problem of obtaining clean drinking water. The open surface water was hardly useable anymore as the sewers all ended up in there, so clean water had to be pumped up from deep ground levels far away from any sources of pollution. Although in the Netherlands there is hardly a spot without access to groundwater at some depth, good quality water is not so easy to access. However it was soon discovered that water of exceptional quality could be harvested in dunes near the coast.

'Water'-companies launched plans to start extensive pumping in these areas without thinking too long about the long-term consequences. Dubois was consulted as professor in Geology and rapidly realized the risk over-exploitation of these resources. Although he agreed with the exploitation of these resources he clearly pointed out that the water in the underground depots originated from rain and not as his opposition supposed from underground waterways seeping through from further inland. Any pumping therefore should never exceed the rain influx as emptying these depots so close to the coast would result in drawing salt water in from the sea which would spoil the aquifers and would render the water undrinkable for a very long time. It also would have drastically changed the ecology of the area and might well have had consequences for the functionality of the dunes as a buffer against the sea, although that was probably not on Dubois' mind at all. But Dubois started campaigning against the water-industry and acquired funds to drill in many places in the Netherlands to investigate how the subsurface hydrology of the Netherlands worked and how it was depending on rain. Though at first ridiculed and fiercely attacked by those who wanted to make money out of this water, he was very much correct in his assumptions and proved it adequately. If the water companies had had their way at the start of the twentieth century, people around the cities of Haarlem and Amsterdam would even now still have had to deal with brackish water. Although the whole issue got substantial newspaper coverage at that time, Dubois nowadays is rarely credited for his contributions on this front. He was however credited literally: he made quit some money examining the underground with his drilling samples throughout the Netherlands.

Even now the dunes at the Dutch coast are still an important source of clean drinking water due to careful management. Most of these areas are now also designated and popular nature reserves of unique ecological value. Whereas *Pithecanthropus* was his most significant contribution to science, mapping the underground hydrology of the Netherlands and thereby (unintentionally) saving the dune-ecology was probably his most important contribution to Dutch society.

## *Kallilimne*

After being settled in again upon his return to the Netherlands and gaining financial security based on his professorship in Amsterdam, he bought himself a piece of land with a large fen on it which was called 'de Bedelaar'. The house he built there and the whole estate are since commonly known by that name, but Dubois in fact called it 'Kallilimne' which translates back from Greek into 'beautiful lake'.

He decided to turn the thirty-eight-hectare property into a theme garden: he wanted to bring back the landscape which must have been there before the Ice-age. The fossil plant material that had been found in the clay pits from Tegelen provided the data on which he based his planting schemes and to acquire them he had to import all kinds of exotic trees and plants.

DUBO3271 Dubois (second from right) next to Jac. P. Thijsse (with hat), the iconic nature protector of the Netherlands, in front of the large fen 'de Bedelaar', on Dubois' estate. (Thijsse proclaimed on that occasion that he would have much preferred to have seen local plants instead of Dubois'exotics.)

The little lake itself was at that time probably already an example of a very rare oligotrophic fen with its own very specific ecology, but that was sacrificed to his greater good, plants and fish were introduced and from a conservation point of view he completely ruined it and turned it into a eutrophic fen in no time. Suggestions that Dubois had some higher ecological landscape goal in mind are largely overrated: The exotic plants he wanted just would not grow without a huge amount of clay, chalk and fertilizer, which he happily applied without any regard to the natural habitat. He closely monitored the changes/devastation he brought on to the indigenous landscape e.g. accidently killing of thousands of fishes in his fen as a consequence of the suppletion of some heavy metal he considered lacking in the ground.

This is not to say that Dubois was not very nature minded. He was involved very early on in the Natuurmonumenten society, the first society for the protection of nature-reserves in the Netherlands and on his estate he was particularly involved in erecting bat-towers to aid bat-populations to expand (but also very much because he thought this would help him get rid of mosquito's). (Voute and Lina treated the history of Dubois' bat-towers extensively in an article in Natuurhistorisch Maandblad 72(9)1983)

Dubois' bat-tower

Nevertheless, Kallilimne was his retreat from the academic west of the Netherlands and he spent most of his money in turning it into his dream estate. After his death it stayed but a short while in the family: during the Second World War his daughter Eugènie was forced to leave her own house and retreated to the estate where she subsequently kept a number of people in hiding from the Germans. But shortly after the war the house and part of the grounds were sold. The larger part of the grounds still remains in the family with a restored shed carrying the name 'Kallilimne' painted on a board by Dubois himself, but the exotic trees that Dubois planted have all withered away except for a very few.

## The book part II: The Peltenburg era

### *Selenka*

C. Peltenburg, director of the bookshop and publishers E.J. Brill in Leiden, upon the 75th anniversary of the company. (Courtesy of Royal Brill, publishers)

After De Stoppelaar had passed away Corneille Peltenburg became the sole director of Brill and he was much more businesslike than his friend. He immediately put the pressure back on Dubois, in particular by pointing out that a new situation was about to develop now Selenka was going to dig in Trinil. We make a small step back in time to introduce this new face: Margarethe Leonore Selenka. She was an extraordinary person in many respects. Born in 1860 she divorced her first husband to whom she married in 1886 to marry again in 1893 with Emil Selenka, who was the widower of her sister. Under his influence she studied paleontology, anthropology and zoology, and worked as his assistant and was an active participator on several scientific trips through the East-Indies that her husband had organized. When he fell ill and had to return to Germany, she successfully took over his leadership of the expedition in Borneo and stayed there several months to study orangutans. Back in Germany she became involved in women's rights activism and women's pacifism movements and she was in fact the first women to launch an international peace protest in 1899, with a petition at the peace-court in The Hague.

Emil Selenka probably at about 1870 whilst in Leiden

Now Emil Selenka, who had been professor of Zoology for some time in Leiden, and she herself were fascinated by Dubois' results in Indonesia and decided to launch an expedition to Trinil and continue the work on the spot where Dubois had finished, to settle the scientific disputes that had arisen around *Pithecanthropus* by gathering more material and more information on the age of the strata. Emil exchanged letters with Dubois and they were in a fairly good spirit. In one letter Emil even asks Dubois to act as a "postillion d'amour" for a "friend" of his who met this nice lady in Celebes, who was now living only three houses away from Dubois in The Hague… So there is very little doubt that they were on good foot and probably met each other whilst the Selenkas visited The Hague on one of her peace missions. Regretfully Emil Selenka died in 1902, and never came to go to Trinil. Margarethe however did not let this misfortune withhold her from her aspirations. She decided to continue the plans, in honour of her husband, and to lead the expedition herself, even if it meant paying for it herself. She managed to pull this off, the expedition took place in 1907-08, in spite of someone opposing her in any way he could: Eugène Dubois.

Dubois did not take the plans too seriously at first, and exchanged friendly letters with Margarethe, but when it actually looked like it was going to happen, he rallied people like Hubrecht, the famous comparative anatomist, (and also befriended to Emil) to persuade her not to go. Hubrecht understood very well why Dubois was opposed to anyone digging in "his" pit, not only would they have a chance of finding something even more nice, but moreover, they might write it up before Dubois would have, and thereby gaining priority on material on which Dubois was the first to find anything. For that was where it was going to hurt the most, the book that had to be finished and until now never had. At the publishers house the Selenka-expedition also did not pass unnoticed, De Stoppelaar already briefly mentions the threat shortly before he dies and his successor Peltenburg fully puts the argument into words: you will lose priority so finish up! Dubois starts fighting the expedition any which way he can, and writes official letters to the minister of Colonial affairs. How can it be that Germans are allowed to dig in our colonies? How can a woman be allowed to head the expedition? This woman in particular, without any proper qualifications? But the government is not at all opposed to the idea of Selenka digging at Trinil. They may well have realized that this could be THE incentive to get Dubois to finally start writing. They in fact support her with everything except money, even Dubois' assistants Kriele and de Winter are put at her disposal, as are his original maps. An election half way these skirmishes results in a new minister at the department of Colonial affairs and new letters of Dubois, but the new minister answers his lengthy letters with the shortest of replies: having read your letters I see no reason to change the policy of my predecessor. Meanwhile Margarethe Selenka keeps exchanging friendly letters with Dubois and remarks that she has followed the elections in the Netherlands with interest. She is not so sure yet whether to be pleased or not with the new minister, but, she remarks, he is a liberal, and liberals are in general good for science.

Margarethe Lenore Selenka-Heinemann

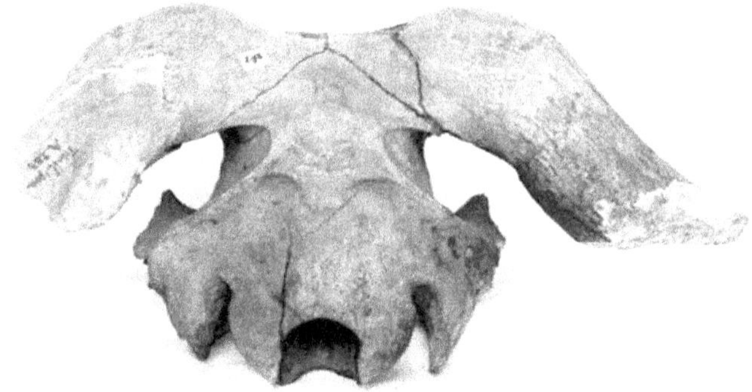

DUBO1243 *Bibos palaesondaicus* DUBOIS, Kendeng fauna.

As it dawns on Dubois that Selenka has too much support and the expedition cannot be halted, Dubois changes his strategy. The Germans have shown some sensitivity to the matter of his of priority and seem open to some deal. But what he (supported by Hubrecht) puts on the negotiation table is naive if not downright offensive. 1: He demands all finds to be sent to Leiden first to be inspected by him. 2: He will immediately return to them any new *Pithecanthropus* finds (which he can promise them easily because he is pretty sure they are not going to find any) 3: Any other material which he needs for comparison to material to which he holds priority should stay in Leiden on an indefinite loan until he has described his material. The Curatorium of the Berlin University responds to this in utmost decent terms: they cannot agree but offer him that as soon as the material has arrived he can come to Berlin on their costs to inspect the material as part of a committee that will decide who is to work on what. Any material other than *Pithecanthropus* finds that are new and that he wants on loan will be transported to Leiden on their costs and can stay in Leiden for a maximum of three years after which it will be transported back to Berlin, again on their costs. Dubois does not agree and does not respond to this offer. He seeks legal advice to start a legal procedure to enforce his demands but is being told that he has no chance at all. When the material arrives in Berlin he awaits (or pretends to await) being invited, but as he never accepted their offer they do not feel obliged to do so, and don't. After inquiring with a 'have you forgotten me' letter they do allow him to come and visit and will allow to loan him this or that, but now only for a short time, as they have already ordered the finds into the museum. Again Dubois seeks legal advice, this time to enforce the offer that he did not accept at first. Again he is being told he stands no chance at all and his legal advisor Groenveldt adds in a last sentence: you better hurry up with your own publications. And so he does. No extensive descriptions though, the bare minimum to gain the priority. In the twelve years since his return he has not written one word about the fauna of Trinil and Kendeng, now he describes most of the new species in two publications, one in Dutch, 10 pages and one in German, 36 pages, with

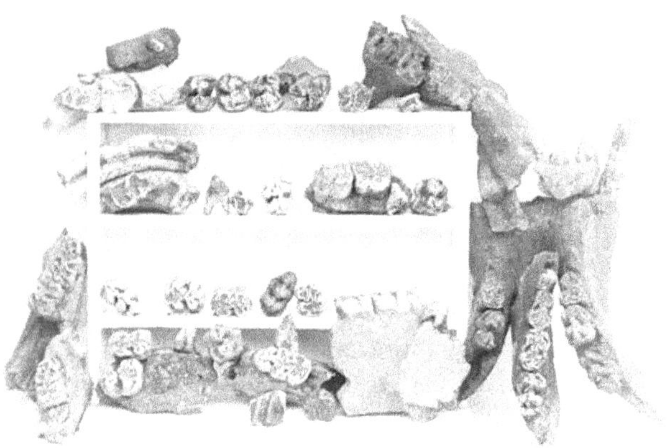

DUBO0827 *Hippopothamus sivajavanicus* DUBOIS, Kendeng fauna.

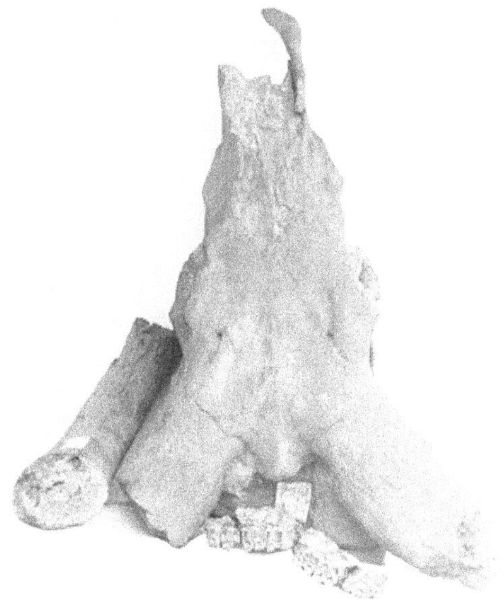

DUBO0970 Skull of *Epileptobos groenveldtii* DUBOIS, part of the Kendeng fauna, named after W.Groenveldt, highly placed civil servant in the colonies and supporter and later also occasional legal advisor of Dubois.

the descriptions so short and in a matter of fact style that it drives taxonomists into despair. As scientific descriptions they are worthless, but they do just contain the bare minimum information to make them valid as species descriptions and thereby ensured his priority. Stremme, who is describing the mammals found in the Selenka-expedition writes in his report: "Whilst I was working on this material in the autumn of 1908 a publication of Eugène Dubois came out,... in which he for the first time gave a description of almost all the mammals of the Kendeng-layers. In Dubois' view they got new names and very short, but to the point characterizations. This work I had to deal with first. But as skeptical as I was, with regard to Dubois' new species descriptions, I have to admit now that I agree with almost all of his conclusions."

Although the Selenka expedition had been a success, no *Pithecanthropus* was found. Margarethe thought she had some teeth but Dubois dismissed those correctly as much more recent than *Pithecanthropus*. His own publications ensured his priority for almost all species unearthened. Dubois had come out on top. But he must have realized that luck had played a major role; Margarethe might just as well have found a nicer skull than his', which would have put him way back out of the limelight.

## *Finale*

Whilst all the Selenka business is going on Corneille Peltenburg puts pressure on Dubois to finish his book. His first letter to this purpose dates 25th of July 1906 and remains unanswered, the second more pressing letter dated 25th September 1906 urges Dubois to respond to the threat of Selenka. Dubois writes back that he has a deal with Selenka so Brill should not worry. This was, as we know now, not entirely true but this statement buys him some time and some respect of Peltenburg who had not expected this reply. Dubois must really have thought about finishing up but even after all these years there is still the matter of Martin's name on the front cover as Editor-in-Chief of the 'Sammlungen' and his right to view the last version. Dubois writes in 1907 to Hubrecht and Weber for support to have Martin's name removed and to ensure Martin has no possible influence on the contents.

Ambrosius Hubrecht

Hubrecht and Weber reply shortly that they see no problem in Martin's name being on the cover and that Dubois should not bother about that. He can't disregard them, they are too important to lose as supporters, and no doubt their refusal has played a part in Dubois' decision to publish his short notes on the Kendeng fauna in which he secured his priority on the new species as we seen on the previous pages. Surely that subject ought to have been a major part of his book, but now they are published elsewhere the publication date of the book is further away than ever. In 1907 Dubois seems to need money and asks Brill to pay for the expenses of the photographs that have been made for the book such as the Stoppelaar has promised him. Peltenburg does not fall for that and the copybooks of Brill proof their worth: he produces the letter of De Stoppelaar that clearly states that money is only to be paid after the government money has come in. Dubois has no claim and with this action has lost a lot of credibility. Peltenburg puts the screws on and demands a definitive date when the work shall be finished. When Dubois does not adequately reply Peltenburg makes clear that as far as he is concerned Dubois has two options: finish the book or pay for the expenses incurred so far. In October 1908 Dubois promises that he will do his best to finish but he doesn't do anything. Peltenburg tries again three years later, although in his mind he must have already given up and write off the sums due to the firm. Another few letters of good intention are written by Dubois, the last one being the 1913 letter with which we started this story earlier. Peltenburg launches his final attack, threatening as De Stoppelaar did to publish everything that is in his possession without Dubois' cooperation. Dubois thereon (without literally stating it in so many words) threatens Peltenburg to go to court should he do so. Dubois complains that Peltenburg does not want to discuss the matter in private, for then it would all have been resolved a long time ago (…) and he ends that he will next year come with a major part to be published. Peltenburg however throws the towel and does not respond…

Corneille Peltenburg behind the desk at the Brill offices which he held as sole director since De Stoppelaar's death in 1906 until shortly before his own death in 1934 (Courtesy of Royal Brill, publishers).

The finale of this drama comes rather grotesquely two years later in 1915. Dubois receives an advertisement of Brill in which he sees two articles of his own hand offered for the very low price of 60 cents. Brill had in those days next to their publishing activities also their own book shop and Dubois had a long time ago supplied some publications to De Stoppelaar to be sold in their shop. But now the price had gone down so much he decides to inquire after the cause of this and sends a letter to the firm, maybe not even realizing that this letter would end up on Peltenburg's desk. Peltenburg replies:

*Most learned sir,*
*In immediate reply to your honoured letter of yesterday, we inform you that the works you refer to are antiquarian examples and as such are not part of the supplies we have in our depot that belong to you and which are for sale for the prices of 2 guilders and 25 cents for the English article and 90 cents for the German article. In store are still 271 exemplars of the German and 109 of the English article which we will be happy to return to you should you demand so.*

*On arrival of your letter we anticipated to hear from you that the manuscript, on which we have been waiting for so many years now, would finally be ready. You should be able to understand the disappointment not receiving any mention of it at all.*
*Sincerely*
*C. Peltenburg*

I fear I have not been able to sufficiently express into English the utter contempt that Peltenburg must have felt for Dubois, and which is still dripping from the page of this Dutch letter…

DUBO0777 Henry Fairfield Osborn

In 1924 Dubois finally publishes illustrations of *Pithecanthropus* after enormous pressure from scholars worldwide headed by Henry Fairfield Osborn, as has been extensively documented by Bert Theunissen and Pat Shipman. Some of these illustrations are of the ones originally intended for his book and were amongst the plates of which Brill had 600 in print: not very elegant to say the least. Peltenburg is than still in office at Brill, but no response of his side to these publications was given that we know off. He remains director until shortly before his death in 1934 at the age of 80 years.

Until his death Dubois kept defending the special position he thought his *Pithecanthropus* had and he debunked every other *erectus* find. Not the Peking man and not even the skulls from Indonesia, which were found by Von Koenigswald, were to be considered *Pithecanthropus* in Dubois' eyes and he was very wrong and unjust in his judgment about them and the people who discovered them. He just could not see how the other finds strengthened his own find as being the first found missing link in the evolutionary tree of the humanoids, just for want of being the only one. Bert Theunissen showed us that his theory on saltatory cephalization more or less forced him in the end to announce that his *Pithecanthropus* was more gibbon like. His adversaries, Von Koenigswald in particular, used this to declare that Dubois had now completely lost his marbles and that he denounced the humanoid status of his own missing link. Meanwhile the Second World War had started, people had other concerns, Dubois had moved out of his house in Haarlem to completely retreat on his estate where he died shortly after, complaining of the cold in his last letters to his daughter Eugènie.

DUBO0139 Weidenreich (left) and Von Koenigswald (right)

# Epilogue

*The Dubois collection in the Netherlands*

Father Bernsen (collection Geological Museum Artis)

Since 1894 the fossils were packed in 400 crates and sent to the Netherlands where they were kept at the 'Rijksmuseum van Natuurlijke Historie', Leiden. The idea was to store them temporarily in the gallery of the building at Rapenburg, where the 'Rijksmuseum van Natuurlijke Historie' was housed. This gallery was in earlier times used for the geological collections, but after the separation of Zoology and Geology it was empty. This gallery was appointed by the Ministry of Internal Affairs for storage of the collection. However, the director of the museum had the space already occupied. So, as first emergency storage the collection was stored in a house on the grounds of the Vreewijk estate at the Witte Singel. However the owner was afraid that the building could collapse under the weight of the collection and then there was decided to store the collection in 'het koetshuis' at the Doelensteeg that belonged to notary Obreen. In 1915 the collection moved to the former building of the Rijksmuseum van Natuurlijke Historie, and in 1919 it moved on to Breestraat 22. In 1925 it moved to the Psychiatry building of the Academic Hospital, and in 1930 to the Boerhave Laboratorium.

Top: DUBO0096 *Panthera tigris* described by Brongersma from the Dubois collection, below Dubois and Brongersma in 1935

From then on father Sanctes Bernsen was appointed assistant to Dubois in which capacity he catalogued almost the entire collection. The tragic death of Bernsen, who died on the 5$^{th}$ of June 1932 of acute stomach bleeding, has been attributed to the stress that Dubois had caused him. Proof of these accusations will never be obtained but we can be sure that the stress Dubois caused the people around him would not have helped cure anybody with a weak constitution.

Bernsen was succeeded by Miss Margaretha Sanders, who later became the wife of Leo Daniël Brongersma who himself became an assistant to Dubois from 1933 until Dubois' death and subsequently remained curator of the Dubois collection until 1946. Brongersma described the Carnivora of the Dubois Collection.

Karl Martin aged 75 (1926)

In 1934 Dubois placed his definitive coup to ensure that his collection would stay within the realm of natural history as opposed to the geology museum of Martin. Not that Martin was still in office. The new director of the Geology Museum, Escher, tried his best to get/keep the Dubois collection, but seems not to have been at the base of this battle. The archives preserve a letter of Ralph von Koenigswald in which he makes plausible that it all started with a dispute between Dubois and Isaäk van der Vlerk, conservator at the Geological Museum. Martin might well have been pulling the strings on the background but whatever the cause was, Dubois secretly initiated a petition of 25 colleagues to persuade the responsible Minister to definitively bring the whole collection into the Natural History Museum. As was to be expected this drama took place with all kinds of possible subplots, people breaking the secrecy, people pointing fingers at each other, animosity between the geologists on the one side and zoologists/anatomists on the other side, but when the smoke had cleared Dubois had his way. (Much, much later both musea fused, probably causing over fifty odd players in this pre-tv-soap-opera to turn over in their graves)

During 1939 and 1940 the collection was moved to the main building of the Rijksmuseum van Natuurlijke Historie. In 1946 Dirk A. Hooijer became curator of the Dubois collection, and described the fossils one by one in a series of papers, like the Proboscideans.

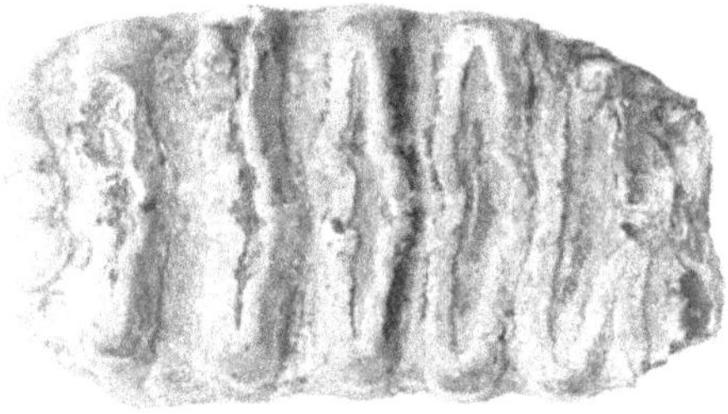

DUBO0178 *Stegodon* molar from the Dubois collection described by Hooijer

Hooijer's taxonomical work enabled the study of the faunal composition of the hominid 'type localities'. Whereas he did not use the fossils in a biostratigraphic way, this became now possible and was the starting point for a new look at the old collection.

Dirk A. Hooijer

Hooijer retired in 1979 to be succeeded by John de Vos. In 1979 the collection was partly stored in a room of the building of the Douzastraat belonging to the Rijksmuseum van Natuurlijke Historie and partly in the store-rooms of the dependence of the Heerengracht at Leiden. In 1981 the whole collection moved to the former 'Rijkszuivelstation' in the Vreewijkstraat, where De Vos also was housed. In 1997 the collection moved to the 16th floor of the tower of what now is Naturalis.

Although John de Vos is co-author of this book, this part is solely written by Paul Albers as it needs to be stated somewhere that John de Vos has been the driving force on the background of our current appreciation for Dubois. He very much initiated the biographies of both Bert Theunissen and Pat Shipman. In his own research he picked up on the biostratigraphy where Hooijer left off, and identified the different faunas from the different locations where Dubois had been sampling. The archives, which were hardly more than a bunch of cardboard boxes when he started have since slowly but surely been inventorized and are now on the brink of being completely disclosed and publicly accessible on the internet. In the course of this book more than 2300 photo negatives have been digitized.

The future will still hold new views on Dubois' remarkable life. In particular Dubois' geological contributions have not yet received much interest and would be a nice project for someone with an interest in geology and the history of science.

## Appendix I: The Brill correspondence

The copybooks from the Brill archives are now stored in the library depots of the Amsterdam University, and are safe now, but they have not always been stored with the care needed to ensure future preservation. Some have suffered water damage and one is lost entirely. The lost one holds two relevant letters, which luckily can be reconstructed as from those letters the replies of Dubois still exist in drafts in the Dubois archives. Presented here are the English translations, essentially unabridged, but as some parts or words are undecipherable, translation is sometimes hampered and for the sake of readability I (Paul) have not always given account of uncertainties. The Dutch originals will in the near future be accessible on Naturalis website. *Indecipherable/unclear* areas are placed against a grey background and when with text this means it says what I *think* is there. Dots indicate missing word(s).

One extra remark needs to be made about the letter beginnings. In English it is very much a custom to almost in every case start your letter with Dear Sir and avoid all frivolities indicating rank, status or the pretence of those, as well as the mockery. I did choose to at least try and incorporate an indication of this in the translation, though I realize it must seem very odd to anyone with an English background.

---

*Very learned sir, Dr E. Dubois*
*The Hague*

*The costs of your reprints are for your 7 ...... and plate f13,25*
*post package -,40*
*f13,65*

*May I politely request this sum to be sent to me*

*Regards,*
*Your*
*Leiden Fr. de Stoppelaar*
*8/9 '96*

*----------*

*Telegram 15-9-1897*
*200 or 250 copies would cost 55 gulden per plate, to make drawings or clichés and typesetting are the most costly*
*Stoppelaar*

(Answer to telegram Dubois: What would the costs be per plate for 200 or 250 copies under otherwise equal conditions)

*----------*

*{28-9-1897}*

*Mister Dubois*

*Attached our supplement. I hope I have understood your intentions correctly. Should this not be the case, please send it back to me with your annotations. To your account, which elucidates the matter at hand very well, I have no objections whatsoever.*

*Included in the price of f 60,- is f 5,- to be restituted to you by us after receipt of the subsidy for every plate; and f 2,- idem, idem for every photograph for the clichés.*

*I have no further remarks. Should you need me for help, please let me know: and I shall drop in at the Rapenburg, when you are there.*

*----------*

Telegram 8-10-1897

Question is not quite clear to prof Martin nor to me. Verbal explanation desired. Impossible to already decide on the number of deliveries.

Stoppelaar

(Possible answer to telegram Dubois: Please send me the received corrections [by] Buttikofe. Will you work tomorrow)

———————————

6 nov

Highly esteemed Mr Dubois

Thank you for your letter that just arrived. I am happy the matter is progressing ........ I was convinced nothing ............. have asked, but when your response did not arrive for some time, I feared Mr Bach had asked a contract, with an extensive calculation.

Now my fear has proved to be unfounded. I hope we will be able to go to The Hague soon, when your cold has cured, which won't be easy with this weather. ......... with kind greetings

Regards
your
Fr. de Stoppelaar

———————————

*The Hague. 11.XI.97.*

*Highly esteemed Mr de Stoppelaar,*

*Hereby I send to you, attached , as official record, proof of authorization by the Minister of Colonies for the publication we have discussed repeatedly. The matter is now completely arranged.*

*I am still very busy experimenting with photography, and start to succeed quite well. For the moment my own house will suit me best for that purpose, to which my cold, that still has not left me, is restricting me anyway .*

*I hope you are well and will stay, after polite greetings also to Mr Van Oort and Mr Pleyte, with the highest esteem,*

*Your loyal servant,*
*Eug. Dubois.*

———————————

*The Hague, the 10th of November 1897*

*I am honoured to announce to you, that the Minister of Colonies has authorized me, as noted in Missive dd 9 November j.l. Lett A[l] N[o] 9, to compile the descriptions of fossils in a book to be printed by your firm for the amount of max f6350,-, to be divided over three years.*

*At first in the year 1898, after the Indonesian Account for that year has been approved, your firm can draw on the amount of f2000,- by handing over a declaration in twofold.*

*Equally for the year 1899 the amount of f2000,- will be made available, whilst the rest will be paid out after the complete work has been delivered, but not before 1900.*

*Legally signed, Eug. Dubois*

*To*
*The bookshop and publisher*
*previously E.J.Brill*
*at*
*Leiden*

———————————

Leiden 12 Nov. 1897

Very learned Sir, Dr Eugène Dubois,
The Hague

Highly esteemed Sir,

Very gratefully I acknowledge the receipt of the favourable decision of the Minister of Colonies. We can now start our venture. When you are in Leiden at the Rapenburg buildings, I'll be happy to pass by, to make some definitive arrangements, will you please announce your presence by postcard?

Friendly greeting, with high esteem

Your

Fr. de Stoppelaar

———————————

Most learned Sir,
Without wanting to hasten you I'd be pleased to hear from you when we can start with the drawings for your book. I am asking you this also in connection to the planning of... ... .........

Believe me greeting with pleasure and highly esteeming,
your

Fr de Stoppelaar

Leiden
4 Jan '98 {1898}

———————————

15/3 8 {1898}

Highly esteemed Mr Dubois,

With regard to my journey to Germany - which I have set for now on the 20th of April - I'd like to receive notice, whether I could before that time count on the ........photographs and further information needed to discuss the reproductions that are to be made. If you prefer to add your comments in person I could briefly pass by on Sunday morning at your home in The Hague, with friendly greetings,

your

Fr. de Stoppelaar

———————————

28/4 8 {1898}
Highly esteemed Mr Dubois,

I regret also not to have been present when you visited here. I believe I have succeeded well on my trip to Leipzig. I have seen at Meisenbach, Riffarth & Co just ...... ...........geolog. work, that was also produced by them, and seemed excellent to me.
He is prepared to send his photographer with camera to Leiden, if necessary, and given everything is thus prepared, that he can subsequently record all plates. He assumed to be able to make 10 to 20 recordings a day. The price as a consequence will increase substantially, but in return we will get ....precise work. The .... .... ..... can wait in my opinion until they have received a proof and the ...... negatives.
Of all as sharp and as ....... ....... .... increasing the quantity... went well in all cases ....... photograph at the original size.
Once the photographs have been prepared he can deliver everything within six weeks.
We need to confer how to address this. It is reasonable to expect galley proofs of the negatives you have sent in the course of next week. There are no further remarks to be made. The plates, as we discussed, I have all explained to him, and he understood perfectly.
I will tell you more about it when you visit Leiden. Meisenbach will wait ........ first the .......

Believe me please, with high esteem

your
Fr de Stoppelaar

———————————

24/2 1899
Highly esteemed Mr Dubois,

As soon as we have received your text we will immediately start typesetting and as far as we are concerned this can be ready fairly quickly. The letter you intended seen on Prof. Martin's ........ is at the moment very much occupied with different work. I could in the meantime take a certain amount away from those works to have your oration typeset, but only on the condition that it will be printed very soon.

I do dare to suggest to you to choose another or ........ ...... new letter, which we have used to print amongst others the speech of Prof. Veih, which I am sending you under ........ This is als a very fine and apt letter for this purpose.
I gladfully leave the decision to you.
I regret not having known, that you have delivered your oration last Monday, I would certainly have come if I had. I only heard in the afternoon; the gentlemen Schuchat Knol and my son, with whom I had diner at my brother-in-law Lutap, had been there and were still very impressed by the many and large numbers you have summed up - I hope I will now be able to read it

Greetings from the heart and with high esteem

Your
Fr. de Stoppelaar.

---

Leiden $^{11}/_{1}$ '900
Highly esteemed Mr Dubois,
This afternoon I had the opportunity to speak to Prof. Martin, who has returned from Berlin for a couple of days and discuss ......... Of the text. The conlusion was, that Prof. M. for several reasons can<u>not</u> agree, to have only "Sammlungen" mentioned without his name there as editor, because the volumes are always cited in full in geological publications with his name present, and are as such known everywhere.
I have to agree with him on this that this is indeed the case with other volumes also from other societies.
Prof. Martin thus claims for your work to be excepted in the "Sammlungen" that the title will remain unchanged and that he will get to see the final version, to ensure uniformity, ... the editorial board, obliging him not to change anything with regard to the contents, even if he would not agree with it; You are thus completely free in that respect.

If that is not to your liking, it will not reflect on the friendship between you, but in that case Prof. Martin must object to it being published in the "Sammlungen". It should then appear as a separate publication. You will have to discuss this with the Ministry of Colonies and get it approved. Prof. M. from his side will not take the matter up with the Ministry, nor cause any trouble.

- - - - - - - - - - - - -

To you the decision how to deal with this. Not accept, the compromising proposition I made, of Prof. M., for reasons mentioned above, being that the book will be reckoned as having been written under his editorship.
As far as I am personally concerned, it is equal to me.
Whatever decision you will make; I will put up with it, be it that I will

have to deal with a number of clients that have already subscribed to the "Sammlungen" in anticipation of your submission.

I hope that we now soon can start printing.

From the ministry ... ......... ... of .... Mr Loudon ....... wrote me that His Excellency is by no means going to pay anything as long as not 1/3 has been found okay..... the Ministry for........ So on that we also have to wait patiently. Let's hope 1901 will create the opportunity to ......... ...... ... will go, and the ..... ...... ... ..... ............ With fr. greeting signed

Your

Fr. de Stoppelaar

-----------

Leiden 18/1 '00

Highly esteemed Mr Dubois,

The day before last Prof. Martin came to visit me to inform me on the outcome of the talk between the two of you; I ............ you me in as it was decided, and I promised Prof. M. to write to you again, that as to the matter of the proofs, you have misunderstood my writing. What I intended with the final version was the fully cleaned up definitive proof corrected by you. This is subsequently being sent to Prof. M.; he writes his remarks on them. Sends it to me, and I send it to you. So you see it again, and by you the "go ahead" is given without Prof. M gets to see it again.

So you have the last word in every way. Yesterday I was absent, and did not have - also because two of my employees have fallen ill - any opportunity to write this to you. I regret this now very much, because I am currently surprised by your message that you at last have decided to have your book published separately.

The matter is, as you know, equal to me. As Publisher it does not matter to me whether it will be in the "Sammlungen" or not. This book will find its way. That I am writing again about it, is because of the promise I made to Prof. M., as I stated at the start of this letter, and the remorse that my neglect might have caused your changed opinion. May I add as my private opinion, that you, in my eyes, overestimate the interferences of Prof. M. in this matter. I have been dealing with him for over twenty years now in all kinds of matters, often also under difficult circumstances. Often we have had differences of opinion, as is only common, but it has never influenced our friendship, because as far as my experience goes, I have come to know him as honest and true, putting his business ..... ...... , and what is of particularly great value, he always keeps his agreements. If Prof. M. has given you written and verbal affirmation that he will not make things difficult for you, nor abuse his position as editor, then you can believe this without a second thought.

It is my intimate conviction — be it very personal — that you, working together with Prof. M. will be saying that this cooperation has become one that will leave the fondest of memories. ..... Prof. M. — he may be

a bit excessively accurate in some cases — ........ stays the same, is pleased to help, without a hidden agenda.
This is what I felt I had to write. .......... I would not have, then I have got your trust, which I hope to keep. However the matter will be decided, it will not harm our cooperation either way. Believe me truly;greetings, also to your wife .

Your
Fr. de Stoppelaar

———————————

Leiden , 20/3 1900

Highly esteemed Mr Dubois,

Don't think ill of me when I come to bother you with the question whether we can soon expect the next part of your work. I ask this keeping in mind the Assurance given to Mr Loudon,that the first volume of your work would be ready at the exhibition in Paris. As this is being opened at the 14th of April, and according to messages in the newspapers one will not be allowed to add new items after that date. If there are chief reasons or causes, that will not allow you to finish the First volume, it seems to me that it would be good that Mr Loudon is being notified, either by you or by me

Believe me truly, with friendly greetings,

Your
Fr. de Stoppelaar

———————————

Leiden 18 May 1900

Highly esteemed Mr Dubois,

This morning I returned to Berlin from my trip to Leipzig. Your letter has thus reached me in good order; I was happy to find it coincide with my presence there. What turned out to be the case? I was speaking to the Head-Chief (pleonasms like one sees nowhere) in the Berlin House on the "Meise", and heard that Mr. Rig........... left the Leipzig house months ago already; ...... ........ (this.... ..........) could be found rather irregularly, and e.g. the last of your proofs had never reached the studios in Berlin. Mr Spiess apologizes for this and requested me to confer his apologies also to you. I will tell you more when I see you. They even did not tell him about ...... , that Rig.... ...... had brought you. I hope now that the drawing has not already been sent to you: if this is the case, please inform me directly: I'll request him to send it to Berlin. Maybe ........ - if their questions don't bother you too much - you might yourself .... .... Heer Spiess, or simply write to the chef of Meisenbach, Riffarth & Co when you send the plates and explain to him what is going wrong. Mr Spiess has promised me extensively that he would deliver them tadellos and as genau wie möglich[5].

I am sorry for the trouble this has caused, but nobody is to blame. At least we now have the assurance that everything will be all right, because Mr Spiess repeated to me with stress, that he hoped that we would not be put off by the incomprehensible behaviour of Rig..., and that the time that had

---

[5] neat and as accurate as possible

been lost, would be made up by them.

Will you send him all your remarks, then he will answer. He will have everything related brought over from Leipzig.

With fr. greetings, and..... , with high esteem

Your

Fr. de Stoppelaar

———————————

Leiden 20/4 1901

Esteemed Mr Dubois

Monday evening I could visit Haarlem briefly to talk with you about your book. Would that be convenient to you, and at what time? I could take the train at 5.40 = 6 hours or 6.{0}7 = 6.27 Haarlem, hence about 7 o'clock at your place. Should you wish another day, fine with me, but not on Tuesday or Thursday.

Monday would suit me best

Greeting with high esteem

Your Fr de Stoppelaar

During the day anywhere next week would suit me

———————————

Leiden, 8 June 1901

Highly esteemed Mr Dubois,

I forgot to take up the matter of the expenses in Berlin, but I don't consider it worthwhile to even try get them reduced, as I could not agree with myself in what way I should present it. It is difficult to write to a chef that his employees have either been lazy or not up to their jobs; and that <u>that</u> has been the cause of the expenses increasing. My plans to go to Berlin one of these days have now become a certainty. There is an international publishers congress in Leipzig starting next Sunday: I will be going there. The congress...... [closed] in Berlin, where we will arrive at the 14th of June and I ....... to go to Spiess and .......... ..... go about the business. It would be of great service to me if either you or Mrs. [Dubois] who is also well informed, would present me a list, before Friday(when I will be leaving) with all the points that we have discussed, and of which I still remember for the most part, what it is all about. I promise you , as is obvious, that I will use it discreetly, as is in the interest of the cause. To cut off a couple of hundred Marks , only to pay the same amount to somebody else, thereupon risking, not to be served well is not the diplomatic way.

I'm rejoicing, that we now will be able to start soon. With you I trust, that both yours and our success will be good.

If there is anything else I can bring up in Berlin: please let me know.

Either way I hope that we will succeed with the continuation in Amsterdam.

Believe me truly, with fr greetings, also to your wife

With high esteem your

Fr de Stoppelaar

———————————

Leiden, 8 Oct. 1901

Highly esteemed Mr Dubois

As I was absent yesterday, this is my First opportunity to answer your letter dd 5/10 last - I regret very much that your wife has travelled to

Leiden in vain, the more as my wife and daughter also happened to be just out.
I have written a pressing letter to Meisenbach. I hope he will now keep his promises. I can imagine that you are getting worried about the way things are going, but if I were you I would not get agitated too much. As all worldly things, it will be all right. I just hope that you will now find the time, ........ [and] and the opportunity to give what you got. By always looking for better the good might stay hidden, and that would, after all that has been done already, be a shame.
........I regret that you haven't sent the drawings or rather photographs to Berlin. I wrote him, as we agreed, they would follow within days. Now he will say; I did not hurry as you did not send what you promised. If I'm correct, you had planned to have this be part of the volume that is to appear: if this is so, then I think it is justified, to send this also to him (Meisenbach) as soon as possible. He will help us now quickly, and then he won't have an excuse not to.- I understand very well why you did <u>not</u> do it, but I think it better, that he is to receive them as quickly as possible with your instructions, to not give him any motive for further delay.
Once we have this First part ready, I promise I will do the impossible to get things done here in Holland. When we started all this there was no studio that could satisfy your demands: now there is.-

Let's for the moment just carry on with Meisenbach, but without giving them reason to complain, so we can get rid of them as soon as possible. If you help us in this, everything will be fine. You do have to trust me, I will hurry up very much, if we only got a partial volume ready for the ministry, all will work out: Mr Loudon is willing enough. With fr. greetings, also to your wife
Truly with high esteem, your
Fr de Stoppelaar

———————————

Leiden, 10/10 1901
Highly esteemed Mr Dubois,
This morning I received letterI from Meisenbach, followed at 1 o'clock by letter 2. So there is some delay from their part. I will notify you when the post package arrives (I expect within a few days). I hope you have sent the other material or will do so now ... "Frappez le fer pendant qu'il est rouge" - they are now in good .........., let's keep them that way by answering soon
With high esteem, greeting
your
Fr de Stoppelaar

———————————

Leiden 11 oct. 1901
Highly esteemed Mr Dubois,
As you have read from the letter, the post package with the plates from Meisenbach has been sent - So you will be receiving it within a couple of days. After that you can establish which negatives still need to be sent to them.

As far as the line drawings are concerned: please be so good as to send them to me in Leiden the first time. I will then see what method of reproduction is best suited. Subsequently they can be sent directly to Amsterdam, but the first [version] I prefer to do myself. Where the clichés in between the text are concerned: these I would also prefer to see myself first. If necessary I will take them to Amsterdam, and have proofs sent to you from there.
It seems, I think, best to agree, that unless we say otherwise, everything will go passed my office. That way I will stay better informed, and it will never cost more than one day.
With high esteem and greetings
your
Fr de Stoppelaar

———————————

Leiden 18/10 1901
Highly esteemed Mr Dubois,
In response to both of your letters I have this morning -I was absent yesterday - written a long letter to Meisenbach, and dotted the i's. I hope it will help. ...... I have asked now, to, as suggested by Mr [Winckelmans] , to send you prints of everything you left in their hands. I will keep you appraised and send you written account of our correspondence. I have no secrets for you in this respect: I prefer to deal with this in the open. The drawings I am sending to Roeloffsen & Hübner in Amsterdam, t...................ical reproduction: as soon as you have further drawings, I'll be pleased to .......... Tuesday I will be visiting Amsterdam, and forementioned Gentlemen , and tell them they can be expecting more work from you. Today I have no time, as I will be leaving for Paris for business for a couple of days on Monday, when I return I'll be pleased to take you to this firm some day, to acquaint you with them. Believe me truly, greetings with high esteem
your
Fr de Stoppelaar

———————————

Leiden 13 Nov 1901
Highly esteemed Mr Dubois,
Your letter dated $^{12}/_{XI}$ last has arrived in good order with the enclosed drawings. Given the needed accuracy for their reproduction, I do not consider zincography desirable.
When I have two zincos made- one for [black] and one for red accurate printing is surely possible, but we need to take into account the paper, which stretches or shrinks due to temperature, minimal as it may be, which would just spoil the correct distances. Therefore I gave the drawings to the lithographer, who completely controls accurate reproduction, and he can , what you have drawn in red, by giving it a different width, create a clear difference. I have put to him the importance of being very precise, and we will get neater lines than if he would used zincography: [with] a tiny width still more beautiful than any other reproduction.
I will write to Riffarth again.
Excuse just these few words. I am in the middle of the mess of

reconstructing my house. Greetings truly

your ......

Fr. de Stoppelaar.

———————————

Leiden, 31/12 1901

Highly esteemed Mr Dubois

I will not let this year end before having replied to your letter of 23/XII, and your reminder of 30/XII. But first of all my best wishes for you and your beloved for the soon arriving 1902.

As to your proposition, to have one plate reproduced by another method: I am prepared to do everything feasible to get as close to the original as possible, but I stand by my opinion that by far the best results are achieved by means of photo engraving. The prints of these are at least much better than by any other photographic method. I am thus convinced of my case that I would propose to you to give me the negative of the plate in question and allow me to make a proof of it at my risk and my costs. The plates in Lavoisier's book are good, but certainly not better than those made for you by Meisenbach & Riffarth. The difference lies in this, that you judge other people's work, I would almost say with a more [lenient] eye than your own. The tiniest of errors or impediment in your plates you wish to have removed, and I agree with you on that , as far as is possible; but I suspect, if you were the author of Lavoisier's book, you would look at those plates with a different eye. Don't think ill of my explanation: it is only human, my experience has brought me to this conviction. I am quite sure, if I had shown your plates, the [proofs]I mean, of Meisenbach & Riffarth to one of your colleagues, they would not have liked them, not being there own work......

Friday afternoon I will be leaving for Germany for about ten days. I might stop over in Berlin? Is there anything I can do there for you?

With greetings, your

Fr de Stoppelaar.

———————————

15 Januari 1902

The most learned Sir

Professor Dr. E. Dubois

Haarlem

Most learned Sir,

As agreed I went ........ with the negative to the firm Roeloffzen, Hübner & Van Santen - building ...... "News of the Day" to discuss the possibility of a good kind of reproduction. According to them the only method for good results is "photo-engraving". The print provided with the negative they considered insufficient due to poor developing. I have now agreed with them that their technical director for zinc-engraving, collotype, photo-engraving etc., a German, a very capable man, will visit you if you such desire, or if that would suit you, you could go visit him some day, after which you could determine your preference.

With the highest esteem remaining

........... ..... [servant]

C Peltenburgh

———————————

Leiden, 28 Nov. 1902
Highly esteemed Mr Dubois,

The crates with the prints of the 24 plates for your book have just arrived. Matters are now becoming serious. I hope you will now be willing and able to cooperate and have a volume ready as soon as possible. Soon we will have to pay for these plates and the expenses, and I am counting on it to be, at least for part of the sum, covered by the subsidy of the Government. I will have the crates opened within the next couple of days and will send you a few and I would like to propose to visit you at the start of next week in Haarlem, to set up a scheme which you and we mutually need to stick to. If this convenes with you, please let me know by postcard which time would suit you best at what day? Every day will be fine with me except for Tuesday, which is fully booked. With friendly greetings to your wife and you,

Yours truly
Fr. de Stoppelaar

———————————

Dear Mr Dubois,

I have just- 8 o'clock - come home from Amsterdam, to find in my office your postcard. Tomorrow, Sunday, I regret will be completely impossible to come and visit you as I will be getting visitors myself. So I will settle for Wednesday next. I'll be arriving by the 9.14 = 9.34 train and will be in Haarlem at about ten so at about half past ten at your place. I had written you directly when I arrived from Berlin. I hope to receive your answer before Wednesday. Did you read the papers this morning ............. something about your work, that about 1/3 ..... .... ready, to be published soon. With fr. greetings, with high esteem your

Fr. de Stoppelaar
Leiden, $^{29}/_{11}$ 1902

———————————

Leiden 14 Januari 1903
Highly esteemed Mr Dubois

Did you receive the prints of Meisenbach and of Roeloffsen, Hübner & v.Santen ? The clichés of the last I mean.

I would so much like to start printing, as ........., so much work is not just .... a loss to us, but I also fear we are getting into trouble with the Ministry. I had hoped that Mr Loudon by soon ............. ..... back ......... And before Tuesday 18 I'd be pleased to have the first print. Help us and yourself. With fr. greetings, pleased with high esteem,

your
Fr. de Stoppelaar

———————————

Leiden $^{16}/_{1}$ 1903
Highly esteemed Mr Dubois,

I have sent Mr Spiess, firm Meisenbach, R & Co a very snappy letter, which will not please him. I do not understand the carelessness of that firm. I have asked them now to return all the negatives to you if they can't deliver all prints. I hope this will resort ........ .
.... Tomorrow........ , .... ............ ...
...........
With fr. Greet. and high esteem

your
Fr, de Stoppelaar

———————————

20 January 3. {1903}
The most learned Sir
Professor Dr. Ed. Dubois
Haarlem

Most learned Sir,
We are pleased to send you the enclosed answer of the firm Meissenbach Riffarth & Co as well as the sketches for Tables 25 & 26.

We expect you will return the letter to us after you have used it. With the highest esteem remaining

your.........
C. Peltenburgh

———————————

Leiden, 18/5 1903
Highly esteemed Mr Dubois,

Although reluctantly, in service of my firm I am obliged to bring up again the matter of the publication of your book. With the closing of the books it became again apparent to me how large the ..... is, to which we have committed ourselves, without any assurance to see anything of that returned, without your cooperation. The Ministry of Colonies will not pay out before it has received a copy. And if I tell you that we have already paid 6000 Marks to the firm Meisenbach, Riffarth & Co, about 3500 guilders, and it has already been a couple of years since we ...... printed, then I am not exaggerating to have a total sum, including the rent, of 4000 Guilders of advanced money.

As I mentioned to you before: "Le mieux est l'ennemi du bien"; it is indeed necessary for you to publicize your results and finds, which you have gathered until now; obviously there will still be gaps, but you will have the opportunity, as we may hope, to fill those in later. The interested parties are allowed to demand to finally get ..... .............. ...... .... .........
Therefore, help us and by that also yourself. Once the first volume is published, the others will surely follow. ..... we are pleased to cooperate. With fr. greet. and high esteem,

your
Fr. de Stoppelaar.

———————————

Leiden, 5/10 1903
Highly esteemed Mr Dubois,

After an absence of more than 5 weeks, which I have spent in Ems for my health, I have returned to my office a couple of days ago. I was hoping amongst other things to find an answer from you, but I regret to have been disappointed. I thought you would have had the opportunity to use the holidays to work on your Standardbook, and that we would now be able to start printing. Would you please be so kind as to inform me, how things are going? I am not just sustaining, as I have written to you earlier, large financial damage, but now also a daily returning grief, of having to give evasive answers or half ones to the question that keeps being asked: when will your book finally appear? One assumes that we, being the publisher, are the ones to know, but the curious situation is that we know nothing of our own publication.

I repeat: help us and yourself

With fr. greetings and high esteem

your

Fr. de Stoppelaar.

---

Leiden, $^{30}/_{10}$ 1903

Most learned sir

Prof. Dr Eug. Dubois

Haarlem

Highly esteemed Sir,

As I had written below the proofs we will be ready with typesetting this week or early next week latest. - At the moment I have no idea what the first part of your book , that we ought to be sending to the Ministry, is going to look like. What is being typeset now is being done so without page numbers - I discussed this also with your wife - , but what to do with the already printed 40 pp., of which some next are already in proof? In other words, in which order are these going to be put?-

We currently hold: 1st these 40 pp. 2nd ± 44 pp. which are currently in print 3rd a considerable collection of plates. - If these are to be assembled into one volume then there will have to be, obviously, a title - which should be easy enough- but also, in my opinion, a preface, explaining your methods, which can be changed at a later stage, but for now will serve its purpose for the Ministry.- Finally, to underline your wonderful plates, a clarification that can be folded out, which could be printed in a smaller letter.

Maybe these propositions are {superfluous}, if so blame that on my ignorance. - The sole intention of this letter is to be armed and ready, when we face the Ministry.-

A separate note from your hand could add some more particulars about how the continuation is going to be, and how you are going to go about it. - If you think a conference is needed I am prepared to come to Haarlem, if I have to, whenever it suits you. In half an hour we can make a lot of arrangements, so I will not take much of your time. With high esteem, greeting, also to your wife

Your

Fr. de Stoppelaar

---

Leiden, $^{4}/_{11}$ 1903

Highly esteemed Mr Dubois,

With regard to the .................. urgency, I request you to notify me if everything can now be printed, what you have signed off for printing ? You have now received proofs of everything we have sent. As soon as you have been in The Hague , all will soon end well. Greeting with high esteem

Your

Fr. de Stoppelaar

---

Leiden 18 Nov. 1903

Highly esteemed Mr Dubois,

I received your postcard this morning, but I don't understand the Ministry' s answer. Is the second term now available without us having to submit anything? We have sped up typesetting and printing, printing these pages without numbers, which means they have to go through the presses again, and now they end up being put next to the other pages and plates. - It seems to me, that we are now entitled, having such a large amount of plates ready, and a large part of the text printed, to go the Ministry ............. , and say, that we have fulfilled an obligation, and we request payment. The interest, that we will be charging, will all be at the account of the country, whilst it should ............ how ......... we too ......., and how much we would like to help, to subtract such an amount of money from our company for just one book.

Once again: please help us by sending us the "missing link" of your Monograph, to make it complete.

Believe me truly greeting with high esteem

Your

Fr. de Stoppelaar

———————————

Leiden 18 Mei 1904

Highly esteemed Mr Dubois,

For several years now, when the books are closed, I have encountered the sum of more than 4000 guilders paid..... the interest, which now has increased again with the reprints of the pages, which last year had to be printed in such a hurry.- As director of our company I can no longer justify this for our board of commisionaries and one way or another this will have to end. - As your wife and you know I have had enough patience , and have truly not been unwilling or untrue, and ............ That is not to be as long as a start can be made with the publication.

With high esteem and greeting

Your

Fr. de Stoppelaar

———————————

Leiden 9 Sept. 1904

Mister Prof. Dr Eugène Dubois
Haarlem

With high esteem,

On several letters of me, though polite, I have lately not even received an answer anymore. I don't think I have done anything to deserve this. I have started my venture with you in good trust. No financial sacrifices have been too large; I have always met your personal desires, but now I am no longer ....... of the company, ..... director I am ................ One way or the other this has to end. .......... are lying without interest in my storage rooms as are plates, that I have paid for years ago. Do take steps to enable us to have the first part of the plates publicized. Enclosed letter and ......... you have said to me once ........ The plates have been ready for ages and are still not being finished. Why? You have ordered them, haven't you? ; and I have taken the risk ..... ....... ......... the firm.

I demand a categorical answer, how is the progress on your book and what am I to expect?

With high esteem

Your

Fr de Stoppelaar.

———————————

Leiden $^{14}/_{10}$ 1904

Highly esteemed Mr Dubois,

I <u>have to</u> write you again, how annoying it will be to you and how unpleasant it is to me: - As director of our firm, I can and may no longer leave ........ .......... I will get into trouble if I do. Therefore I hereby politely notify you, that I will collect the printed plates, as well as the printed text and bind these into a volume, which I will subsequently publish. On the cover I will mention briefly, that ..... ........., through no fault on our side, have delayed this publication, but that we hope, that the author will soon enable us to publish a sequel. Should this not be according to your wishes or if you have another proposition, I will await your answer within 8 days. I am prepared to visit you in Haarlem if you want to treat this matter in person, but this year something <u>must</u> appear. With high esteem and friendly greetings Your

Fr. de Stoppelaar

---

Letter of Lorié to Dubois (filed MM774C-000033-490) in which Lorié completely lets Dubois down when he seeks support in response to the above letter.
Dubois' answer to that (not amused) filed in MM774C-000033-486.

---

Leiden 3 Nov. 1904

Esteemed Mister Dubois,

As Mr. Peltenburg has told you,I did not have an opportunity to answer you yesterday, with regard to the postcard sent to me by Mr Lorié.

It is obvious, ..... , ... ......... ........ , after one ......, that publication of your work about Pithecantr. had become a fact, generally put, that by support of the government Financial difficulties had been lifted, to which means several gentlemen, e.g. Prof. Martin, gave up a part of his subsidy, and the Cooperation for Advancement of Scientific Research in our Colonies, have cooperated - we have spoken often about the publication of your book, and I have been contacted too, being the publisher, if there was progress or not. - Whilst in the first years I could say: we are busy; the plates are in production whilst printing of the text has started, I had to explain later - as the book did not appear - that Prof. Dubois was otherwise occupied, as duely noted by the newspapers, by working on other projects, and I will no doubt have added - which I don't remember anymore - , that this saddened me: "this costs me a lot of money, that I have already spent on the plates", for thou art also mortal, and who will finish it after you're gone? You will not blame me for that. - Lately I have heard nothing from you. My letters have remained unanswered. And when, quite unexpected, again two Gentlemen came to call on me inquiring on the progress of the publication, I have given the same reply: "The plates are ready, a part of the text has been printed, but I can't get the rest out of Prof. Dubois' hands".

Therefore I wrote to you that I was going to publish what I got so far, hoping to provoke a response from you. I am pleased this has happened. We are now on speaking terms again, everything can still turn out well, if you stick to what you have promised, that you see this book as a part of your life's work. Truly, not a triviality! That people are gossiping about it, you know that as well as I do, but let us prove by the deed that the book will be there, as you have promised now, and all voices will be silent. "Le mieux c'est l'ennemi du bien". Give what you have got; it is such **valuable** material, which you have collected at the expense of large sacrifices, do let it be published under your name, and in your words.

Greeting, with high esteem,

your

Fr. de Stoppelaar

———————————

Leiden 5 Nov. 1904

Highly esteemed Mr Dubois,

I sincerely regret - though I feared it would happen - that the publication of your book has again been discussed at the Ministry with regard to their balance sheet. Apologizing again, because of lack of time, and because of being indisposed (which has surely been a factor lately),without doubt, I agree this with you, ..... ..... not. Therefore something else will have to be written or said. But what and how? I am a bit hesitant advising you on this matter, but I will do it nevertheless, risking that you might take this the wrong way, and start suspecting me too, whilst I - I wish I could convince you on this ...... primarily have your personal interest at heart, before that of your publication ......... Let me speak freely, and add my proposition to it. There is a matter of urgency, and as far as I am concerned, I consider this the ~~best~~ only, and in many respect also the best solution, which will preempt future questions on this subject and will make the matter progress to everyone's satisfaction if you are able , in your turn, to cast off all your suspicions. To business...

Yesterday Prof. Martin came to visit me, also having been subject to inquiry by the Home Office, about the budget of his museum and matters in his care - also about subsidies, that with his approval, for your collection, under his responsibility, are being spent on your servant, present here. - He asked me what the answer should be: he'd be pleased to do everything to avoid difficulties, but ........ ........ ........ being responsible. I ........ ....... ....... that in the course of 1905 most certainly part of your work will be published and I requested him, at least for the balance sheet of 1905 to not arouse any suspicion.

There is nothing he'd rather do than cooperate, and will therefore answer along these lines. Now I know, as does Prof. Martin, that you, for whatever reason ..........., mistrust him - I am calling these serious matters by their names now - and therefore there hasn't been any form of cooperation between the two of you on this subject, whereby you ......... Closely involved;

*you will continue to have the collection at your disposal, until you have finished describing it; only after that if will fall to the Rijks Geolog. Museum under ....... of Prof. Martin. Now no ........ progress has been made, there will be a collision between both interests, the more serious now Berlin also is going to excavate in our colonies; and which might result in that you ............... by that lose priority. This not only annoys Prof. M., but every scientist and every Dutchman. This is the cause - solely - this is being discussed, and you ...... ...... the world is, a tiny falling pebble, becomes a rock, and ends up being a mountain. This cán all be prevented - even now - if you two would start cooperating. Dear Mr Dubois, please believe me. I have known Prof. M. for 30 years; I worked with him a lot, spoken to him frequently, I have needed his help and cooperation on many occasion involving others, and whatever other people might say, he is loyal and honest, of good will, for those who confide in him. This is my case: come to Leiden as soon as possible, let's have a conference the three of us, in which the whole case is discussed openly and which results in a joint report to the Home Office by the both of you, forecasting, that by your mutual cooperation, arrangements will be made, starting in 1905 to publicize regularly. Then - as you will see - everybody will be satisfied; rest will be restored for you, and you will be enabled to calmly work on.*

*I am writing this immediately after receipt of your letter without prior contact to Prof. M. himself, but I am ...... That he will cooperate, although he is very much occupied with all kinds of .......*
*If you can decide to this - and I hope and trust so - send me a telegram or write to me, when you are arriving, and I will deal with Prof. M.*

*Greeting with high esteem*
*Your*
*Fr de Stoppelaar*

---

*Leiden 7 Nov. 1904*

*Esteemed Mr Dubois,*

*Only yesterday evening did I find your reply, which I had feared, although I had a small hope, that you would agree to my proposal. I will have to, though with regret, put up with your decision - Regretfully_, as I am truly convinced, that your judgment on the person in question is wrong. - He does have a high opinion of you and of your knowledge and of the way you have conducted your research and ......... . ..... ......... . Nothing would have pleased him ...... , to see your results published and noted, and your name mentioned with honour. I know this, whatever other people may have told you. But I do understand that I will not be able to convince you and I will therefore never bring this subject up again. I hope one day the day will come, that you will be convinced that he is your best friend. He can and will*

help you by word and deed, and <u>now</u> 7 years have passed without any result, which would have been so much different , if a <u>Cordial</u> <u>.........</u> had existed between the two of you: for your peace and quiet, and for good progress of our cause. But now something has to be done. Would you agree, if I were to go to The Hague on...... to speak to the Gentlemen Dupare and Royer? Before anything else we have to avoid that either unfavourable remarks end up in the account or an interpellation in the House of Commons will take place. I am prepared to do so, but only if it meets your approval, and if you tell me, which promises I can make, also on your behalf.
I could go Wednesday morning.

Greeting with high esteem
Your
Fr. De Stoppelaar

———————————

Highly esteemed Mr Dubois,

In response to both of your letters of the 7th and8th of Nov., for which I thank you, I am happy to say, it pleased me, that for now in The Hague all has been arranged to your satisfaction: this was ...... intention too, when I made ........ proposal to you. You have reached the same goal along another road; and this had more your sympathy, thereby resolving the matter, in which I, on your request, advised you, to the best of my knowledge. The result is good: .... ......... ..... your wishes.

According to our ................ agreement, let us together make your book see the light. I will from my side <u>do</u> everything possible, to make it easy on you. Trust, in the good sense of the word, I will have in you. I <u>know</u>, here is nothing you rather wish for than to see you work finished, but a year passes very quickly, and I hope, that you, in time will meet with the demands, that anybody would expect from you, that you will find opportunity to regularly work on it for a couple of days every week. I will wait for now, and absolutely speak to <u>nobody</u> about it - what I .......... have never done unasked for and what I did say I can justify - and I will just ........, if I haven't heard from you in let's say, three months, to ask you, also taking other work into account, how things are developing. This does not exclude trust; on the contrary, I believe this is the right way to go about it. - Thus we go forward with new courage, with that expectation truly greeting with high esteem,

your
Fr de Stoppelaar

——————————

25 July 1906
The most learned Sir
Prof. Dr. Eug. Dubois
Haarlem

Most learned Sir,

During the meeting of the directors and the board of commissioners of our company, which has just taken place, the President of the commissioners, Professor de Goey, has posed the question, what the progress of your publication was. - The considerable amount of money which this Publisher has advanced for........, has drawn the attention of the Commissary charged with reviewing the account of the company. - As this publication has been dealt with in particular by the late Mr de Stoppelaar, it was not possible for me to adequately answer the question posed to me and I was therefore instructed to make further inquiries.

I would therefore be very pleased if you were to inform me when this publication is due.

Should you wish to discuss the matter with me personally could you please politely respond and inform me on day and hour at which you could welcome me

With the highest esteem remaining

C. Peltenburg Dir.

---

25 September 1906

The most learned sir

Prof. Dr. Eug. Dubois

Haarlem

Most learned Sir,

I regret that my letter to you, dated the 25th of July, has remained unanswered, therefore I allow myself to be so bold as to politely remind you of the question asked therein, when we can expect to start the publication of your work. Since then it has been brought to my attention, that Mrs. Selenka has left for Indonesia, to start scientific explorations in the same area, through which the possibility cannot be excluded that further delay could have the consequence that their results are going to be published before yours, you will understand that I am currently expecting a positive reply.

Meanwhile as always with the highest esteem

Your ........

C. Peltenburg

---

6 October 1906

The most learned Sir

Professor Dr. Eug. Dubois

Haarlem

Most learned Sir,

I have received your honoured letter dated 12th last in good order and have taken the contents to heart. I was very pleased with the announcement, about the agreement you have made with Mrs Selenka.

Whilst I am fully confident, that we will now be going to publish the first volume of your work within the time limit you have set, I am giving you the assurance that from my side I will do my utmost to that cause and have now the honour to be with the highest esteem

Publisher Dr ........

C Peltenburg

---

Letter of Dubois to Hubrecht 24-1-1907 (filed under MM774C-000033-120) confidential In which he seeks support from Hubrecht and Max Weber against Martin being Chief-editor of his publication in the "Sammlungen". Hubrecht also on behalf of Weber answers to him 29-1-1907 (filed under MM774C-000033-122) in so many words that he is seeing ghosts…

---

April 1907

The most learned Sir

Professor Dr. Eug. Dubois

Haarlem

Most learned Sir,

Thanks to your clue I have found the copy of the letter of the late Mr de Stoppelaar, concerning the restitution of the photo-expenses. The letter is dated the 28th of September 1897 and goes as follows:

Attached our supplement. I hope I have understood your intentions correctly. Should this not be the case, please send it back to me with your annotations. To your account, which elucidates the matter at hand very well, I have no objections whatsoever.

Included in the price of f 60,- is f 5,- to be restituted to you by us after the receipt of the subsidy for every plate; and f 2,- idem ,idem for every photograph for the clichés. Thus the matter has been resolved to complete clarity.

After polite greetings I have the honour to be with the highest esteem

Publisher Dr .......
C Peltenburg

———————————

20 September 1907
The most learned Sir
Professor Dr. Eug. Dubois
Haarlem

Most learned Sir,

I am pleased to comply with your request and acknowledge to have received in our depot 272 copies of "Die Klimate der Geologische Vergangenheit", of which 72 are sewn and 200 loose page, and also 109 copies of "The Climates of the geological Past", all sewn.

With the highest esteem remaining, after ...... greeting,

Publisher Dr ...
C. Peltenburg

———————————

17 Juli 1908
The most learned Sir
Professor Dr. Eug. Dubois
45 Zijlweg
Haarlem

Most learned Sir,

In the recent General Meeting of Stock keepers the question was posed again when your book is finally going to be published of which the plates are ready and paid for since years and which are every year are responsible for a considerable loss of interest! Again I was not able to answer ............. this. Last year I was full of cheer as you had personally assured me soon to be able to start printing and even requested my cooperation for a speedy result. Since then I have had no word from you and it is by no means indiscrete to request you now politely but urgently to give me a definitive account about your publication. ..... matter has to come to an end. Our firm cannot affirm this situation and the capital we have invested in this publication has to be returned.

In expectation of your honoured explications remaining

Publisher Dr ....
C Peltenburg

———————————

4 October 1908
The most learned Sir
Professor Dr. Eug. Dubois
45 Zijlweg
Haarlem

Most learned Sir,

You have not answered my letter of the 17th of July , and I really cannot understand why not. We have never been impolite towards you and surely are entitled to be treated in an equally polite way. The matter must be brought to an end whatever which way. If you do not allow us to publish your work then the only solution is that you will repay us the expenses that we made.

Expecting your honoured decision, remaining

With high esteem .............

C Peltenburg

———————————

November 1911
The most learned Sir
Professor Dr. Eug. Dubois
45 Zijlweg
Haarlem

Most learned Sir,

More than three years have passed again since your last letter dated the 15th of October 1908, in which you give me the assurance "that the matter has been forcibly taken in your hands and the work has progressed!" I hope you ....... agree with me, that I am now entitled to a certain explanation stating when it will be ready and the text is to be sent to us. I can no longer satisfy the commissioners of our company with promises, they demand a definitive answer, which I politely and urgently request you to send me.

In expectation of your honoured further response, remaining

.........................

..... ...... ..

C Peltenburg sr.

——————————

*December 15th 1913*

*Dear Sir,*

*although this year I have done the impossible to finish a first and most important part of my publication, which is to be published by you, on the Indonesian collection, I am not been allowed to succeed. One circumstance that impeded me was the obligation to fulfill the wish of the Government, to receive a catalogue of the collection, which has taken much of the time available to me.*

*But also I could not have foreseen the size the field of aforementioned publication has amounted to over the past years. Much needs now to be examined from new points of view.*

*I request your trust, that I will do my utmost, to bring progress into this matter and I believe I can assure you that the long delay has not resulted in a decrease of interest. Interest now is surely larger than it has ever been.*

*In high esteem,*

*your servant,*

*Eug. Dubois*

———————————

*Haarlem, Zijlweg 77, 30 -1-'13*
*To the Firm E.J. Brill,*
*Bookshop and - Publishers, Leiden.*

*Dear Sir,*

*In response to your letter of the 11th last I have the honour to notify you that I cannot agree to the alternatives you propose. The first proposition I can't agree to for reasons best conveyed in person. The second proposition not, because for scientific reasons, both for you and for me, that would be the least desirable that could be chosen. I am certain that such a publication would have no value whatsoever, neither in scientific nor from a mercantile point of view. The good name of your firm would not be served by it, nor would mine, and allow me to point out the following to you.*

*By exercising the second option you would release me from any moral obligation to you, which I am obviously pleased to acknowledge, and I would be free to publish anywhere, how and whenever I wish.*
*The plates published by you would not stop that publication, nor would they profit from it, as it would not be difficult for me to replace them.*

*Also I have to remark, that the plates are not solely your property, as – apart from the intellectual property, which rights I can assert – 1st the photographs were made by me during my stay, for that reason, in Berlin for several months. The costs for that stay were not charged to your account. 2nd You owe me a rather large sum of money for the other expenses I made, without taking into account those which are in the contract we have concerning the publication for the Government.*

*It has always estranged me, that you have never preferred personal contact on this matter over written inquiries. Had you decided to do so, then our difficulties might have long vanished.*

*Meanwhile I can tell you, that I will soon (after the Easter holidays) take the utmost measures, as possible under the given circumstances, to see finished by the end of this year the main part of the meant publication (for which the plates in your possession are intended).*

*With high esteem,*
*Your loyal servant.*
*Dr. ……… Eug. Dubois*

———————————

*Haarlem, Zijlweg 77,*
*15 February 1915*
*To the Firma E.J. Brill,*
*Boekshop and - Publishers, Leiden.*

*Dear Sir, referring to the catalogue No. 71 you have sent me, for which I thank you politely, I have the honour to make you aware of two errors in it concerning the numbers 6949 and 6950 (p. 11). Both small publications were written by me in 1893 and 1895 respectively and not published in Leiden. They were published in Nijmegen and Leipzig and in London, for my account, by intervention of the firm H.C.A. Thieme in Nijmegen. The editions in your possession are my property and were at the time handed to the late Mr de Stoppelaar, previous director of your firm, to be traded for my benefit. In view of this I may be allowed to politely ask whether the price of 60 cts for the sewn English publication is not a bit low.*

*I would be obliged to you if I were to be informed about the mentioned errors and the motive for the low price of the second publication.*

*With high esteem,*
*Your loyal servant,*
*Eug. Dubois*

———————————

16/2 1915

Most learned sir,

In immediate response to your honoured letter of yesterday, we inform you that the works you refer to are antiquarian examples and as such are not part of the supplies we have in our stocks which belong to you and which are for sale for the prices of
2 guilders and 25 cents for the English article and
90 cents for the German article.
In store are still 271 copies of the German and 109 of the English article which we will be happy to return to you should you demand so.

On arrival of your letter we anticipated to hear from you that the manuscript, on which we have been waiting for so many years now, would finally be ready. You should be able to understand the disappointment not receiving any mention of it at all.

Sincerely
C. Peltenburg

The most learned Sir
Professor Dr. Eug. Dubois
Haarlem

# Appendix II: Catalogue of glass negatives and (lantern) positives

The left column states the number under which the negative can be traced in Naturalis, Leiden. The second column if 'empty' indicates an original photograph; '❂' indicates the positive of a photograph of anything printed or drawn, hence a very literal photocopy, often to be used for projection. '◙' indicates the same but than a negative. 'H' indicates that we assume the negative not to be originally from Dubois but from Dirk Hooijer, one of the later curators. 'X' indicates X-ray.

If the description mentions endocast, it concerns a cast made of the inside of a skull, in Dubois' case mainly for the purpose of estimating the brain size. Dubois had a large collection of endocasts, we did not attempt to bring them all to name yet. The website, http://science.naturalis.nl/dubois, will update these data in future.

EDp-year refers to a publication of Eugène Dubois + year in which this picture is present.

| | | |
|---|---|---|
| **DUBO0001** | | Horse (page 32) |
| DUBO0002 | | *Homo mojokertensis* von Koenigswald 1636 |
| DUBO0003 | | *Homo mojokertensis* von Koenigswald 1636 |
| DUBO0004 | ❂ | *Pithecanthropus* brain vs *Homo neanderthalensis* |
| DUBO0005 | ❂ | *Pithecanthropus* brain size |
| DUBO0006 | | *Homo mojokertensis* von Koenigswald 1636 |
| DUBO0007 | | *Homo mojokertensis* von Koenigswald 1636 |
| DUBO0008 | ❂ | *Pithecanthropus* brain size |
| DUBO0009 | ❂ | *Pithecanthropus* brain size |
| DUBO0010 | | *Homo mojokertensis* von Koenigswald 1636 |
| DUBO0011 | | *Homo mojokertensis* von Koenigswald 1636 |
| DUBO0012 | ❂ | endocast of brain |
| DUBO0013 | ❂ | *Pithecanthropus* brain size |
| DUBO0014 | ❂ | *Pithecanthropus* brain size |
| **DUBO**0015 | | *Homo mojokertensis* von Koenigswald 1636 vs recent skull |
| DUBO0016 | | *Homo mojokertensis* von Koenigswald 1636 |
| DUBO0017 | ❂ | *Pithecanthropus* brain size |
| DUBO0018 | | deeply eroded surface in compacta femur of Engelschman plaat, under UV lighting |
| DUBO0019 | ❂ | *Pithecanthropus* brain size |
| DUBO0020 | ❂ | skulls crocodiles |
| DUBO0021 | ❂ | lower aspect skull spalax |
| DUBO0022 | | skull *hystrix* |
| DUBO0023 | | skull *hystrix* |
| DUBO0024 | | skull *hystrix* |
| DUBO0025 | | skull *hystrix* |
| DUBO0026 | | skull *hystrix* |
| DUBO0027 | ❂ | skulls crocodiles |
| DUBO0028 | ❂ | skulls crocodiles |
| DUBO0029 | ❂ | skulls crocodiles |
| DUBO0030 | ❂ | skulls crocodiles |
| DUBO0031 | ❂ | skulls *gavialis* |
| DUBO0032 | ❂ | skulls *gavialis* |
| DUBO0033 | ❂ | skulls *gavialis* |
| DUBO0034 | ❂ | skulls crocodiles |
| DUBO0035 | ❂ | skulls crocodiles |
| DUBO0036 | ❂ | skulls crocodiles |
| DUBO0037 | ❂ | skulls crocodiles |
| DUBO0038 | ❂ | skulls *gavialis* |
| DUBO0039 | ❂ | skulls *gavialis* |
| DUBO0040 | ❂ | skulls *gavialis* |
| DUBO0041 | ❂ | skulls crocodiles |
| DUBO0042 | ❂ | skulls crocodiles |
| DUBO0043 | ❂ | Frog (Field Museum) |
| DUBO0044 | ❂ | skull frog |
| DUBO0045 | ❂ | Giant salamander with eggs. |
| DUBO0046 | ❂ | salamanders fig from Gunther Hecht |
| DUBO0047 | ❂ | mandible of Larval salamander |
| DUBO0048 | ❂ | frogs |
| DUBO0049 | ❂ | *Homo diluvii testis* |
| DUBO0050 | ❂ | mandible tooth of larval |
| DUBO0051 | ❂ | *Eryops megacephalus* Cope |
| DUBO0052 | ❂ | modification of manus |
| DUBO0053 | ❂ | Larval giant salamander |
| DUBO0054 | ❂ | Sperm whale |
| DUBO0055 | ❂ | Whale hunter boat |
| DUBO0056 | ❂ | feeding pattern egg snake |
| DUBO0057 | ❂ | Snake eating egg |

| | | |
|---|---|---|
| DUBO0058 | ۞ | Snake eating egg |
| DUBO0059 | ۞ | Snake eating egg |
| DUBO0060 | ۞ | Snake eating egg |
| DUBO0061 | ۞ | Snake eating egg |
| DUBO0062 | ۞ | limnopithecus jaw |
| DUBO0063 | ۞ | femora |
| DUBO0064 | ۞ | Ape maxillae |
| DUBO0065 | ۞ | mandible sangiran VON KOENIGSWALD |
| DUBO0066 | ۞ | skull sangiran VON KOENIGSWALD |
| DUBO0067 | ۞ | skull sangiran (Paleontologica Sinica) |
| DUBO0068 | ۞ | list of skeletal parts from Trinil Beds |
| DUBO0069 | ۞ | skull cap *Pithecanthropus* vs sangiran |
| DUBO0070 | ۞ | skull humanoid |
| DUBO0071 | ۞ | jaw humanoid? |
| DUBO0072 | ۞ | jaw humanoid? |
| DUBO0073 | | ape skulls (hylobatidae?) |
| DUBO0074 | ۞ | plate 2 Kedung Brubus |
| DUBO0075 | ۞ | plate 9 Trinil site |
| DUBO0076 | ۞ | world map hylobatidae eoceen-recent |
| DUBO0077 | ۞ | Taung child |
| DUBO0078 | ۞ | jaws *parapithecus* |
| DUBO0079 | ۞ | jaws apes |
| DUBO0080 | ۞ | jaws apes |
| DUBO0081 | ۞ | femur |
| DUBO0082 | ۞ | *Sivapithecus* |
| DUBO0083 | ۞ | *Sivapithecus* |
| DUBO0084 | ۞ | Taung child vs Young ape |
| DUBO0085 | ۞ | jaw *Dryopithecus* |
| DUBO0086 | | lower jaws apes |
| DUBO0087 | ۞ | world map *parapithecidae* eoceen-recent |
| DUBO0088 | ۞ | comparison molars from lower jaw |
| DUBO0089 | ۞ | Lower jaws apes |
| DUBO0090 | ۞ | molars *Dryopithecus* |
| DUBO0091 | ۞ | molar *Gigantopithecus* from Hooijer |
| DUBO0092 | ۞ | molars orang from Hooijer |
| DUBO0093 | | femur |
| DUBO0094 | | femur |
| DUBO0095 | | See DUBO0018 |
| **DUBO0096** | ۞ | jaw *Panthera tigris,* Dubois spec. Kedung Brubus (see page 125) |
| DUBO0097 | ۞ | upper jaw Panthera tigris Dubois spec. nr. 1495 |
| DUBO0098 | ۞ | Plate 2 Kedung Brubus |
| DUBO0099 | | Dubois spec. 6484 crocodile |
| DUBO0100 | | crocodile |
| DUBO0101 | | crocodile |
| DUBO0102 | ۞ | anatomic slide through mouthopening |
| DUBO0103 | ۞ | tooth histology Mastodontosaurus |
| DUBO0104 | ۞ | gut of frog, *Rana esculanta* |
| DUBO0105 | ۞ | Medial cut through head of *Eurycea* |
| DUBO0106 | ۞ | gut of *Siren lacerta* |
| DUBO0107 | ۞ | fish?jaws |
| DUBO0108 | ۞ | gut |
| DUBO0109 | ۞ | sex differences *Eurycea* skulls |
| DUBO0110 | ۞ | gut of *Proteus anguineus* |
| DUBO0111 | ۞ | lower jaw sangiran VON KOENIGSWALD |
| DUBO0112 | | crocodile |
| DUBO0113 | | crocodile |
| DUBO0114 | | crocodile |
| DUBO0115 | | crocodile |
| DUBO0116 | | *Homo rhodesiensis* |
| DUBO0117 | | *Homo rhodesiensis* |
| DUBO0118 | | *Homo rhodesiensis* |
| **DUBO0119** | | (nitrate) Eugènie, WA & WB(page 35) |
| **DUBO0120** | | (nitrate) WB, Eugènie & WA(page 35) |
| **DUBO0121** | | (nitrate) Eugènie, WA & WB(page 36) |
| **DUBO0122** | | (nitrate) WA & Eugènie(page 36) |
| **DUBO0123** | | (nitrate) Eugènie(page 36) |
| **DUBO0124** | | view in Haarlem or Leiden? See appendix III. |
| DUBO0125 | ۞ | skull cap Pithecanthropus vs sangiran. |
| DUBO0126 | | molar *Pithecanthropus* Dubois spec., EDp-1924f |
| DUBO0127 | | molar and premolars |
| DUBO0128 | | molar and premolars |
| DUBO0129 | | crocodile |
| DUBO0130 | | crocodile |
| DUBO0131 | X | jaw fragment Kedung Brubus |
| DUBO0132 | | endocast |
| DUBO0133 | ۞ | *Homo soloensis* |
| DUBO0134 | ۞ | drawing of endocast |
| DUBO0135 | ۞ | outline brain |
| DUBO0136 | ۞ | orang utan kiezen |
| DUBO0137 | | endocast |
| DUBO0138 | ۞ | drawing of endocast |
| **DUBO0139** | ۞ | Weidenreich and Von Koenigswald (page 124) |
| DUBO0140 | ۞ | drawing of endocast |
| DUBO0141 | | endocast |
| DUBO0142 | | endocast |
| DUBO0143 | | ? |

DUBO0144 ۞ skull Sangiran
DUBO0145 ۞ molar *Giganthopithecus*
DUBO0146 ۞ *Meganthropus* jaws
DUBO0147 ۞ *Gigantopithecus* teeth
DUBO0148 ۞ skull *Pithecanthropus robustus*
DUBO0149 ۞ molar *Giganthopithecus*
DUBO0150 ۞ *Meganthropus* jaws
DUBO0151 *Glaucopis varians* (bird)
DUBO0152 peaty soil
DUBO0153 *Glaucopis varians* (bird)
DUBO0154 tortoise
DUBO0155 tortoise
DUBO0156 snake
DUBO0157 snake
DUBO0158 tortoise shell
DUBO0159 ۞ skull crocodile
DUBO0160 ۞ skull crocodile
DUBO0161 ۞ *Sivatherium giganteum*
DUBO0162 ۞ tapir
DUBO0163 ۞ orang utan
DUBO0164 H pachypleurosaurus maxilla Winterswijk
DUBO0165 tiger
DUBO0166 tiger
DUBO0167 reconstructie *Pithecanthropus* Dubois 1900 location Raamsteeg
**DUBO0168** river Indonesie (page 16)
DUBO0169 molar *Stegodon*
DUBO0170 molar *Stegodon*
DUBO0171 molar *Stegodon*
DUBO0172 jaw *Stegodon* nr 1646
DUBO0173 molar *Stegodon*
DUBO0174 molar *Stegodon*
DUBO0175 molar *Stegodon*
DUBO0176 molar *Stegodon*
DUBO0177 molar *Stegodon*
**DUBO0178** molar *Stegodon* (page 126)
DUBO0179 molar *Cryptomastodon*
DUBO0180 femur distaal, EDp-1926m
DUBO0181 femur distal, chimpansee, left
DUBO0182 femur distal, gorilla
DUBO0183 femur distal, orang utan
DUBO0184 femur distal
DUBO0185 femur distal, *Hylobates leuciscus*, right
DUBO0186 femur distal, *Hylobates agilis*, right
DUBO0187 femur distal, *Hylobates agilis*, left
DUBO0188 femur ??? , *Hylobates agilis*, right
DUBO0189 femur distal, *Hylobates syndactylus*
DUBO0190 femur distal, *Hylobates syndactylus*
DUBO0191 femur distal, *Hylobates syndactylus*
DUBO0192 femur distal, *Hylobatus syndactylus, left*
DUBO0193 femur distal, *Hylobatus syndactylus, left*
DUBO0194 femur distal, *Papio cynomolgus*, left
DUBO0195 femur distal
DUBO0196 femur distal, *Macaca cynomolgus*, left
DUBO0197 femur distal, *Macaca cynomolgus*, left
DUBO0198 femur distal, *Ateles paniscus*, right
DUBO0199 femur distal, *Ateles paniscus*,
DUBO0200 femur distal, *Semiopithecus maurus*, left
DUBO0201 femur distal, *Semiopithecus maurus* Pandan ♂
DUBO0202 femur distal, *Macaca cynomolgus* ♂
DUBO0203 femur distal, *Homo sapiens*, from Java
DUBO0204 femur distal, *Homo sapiens*, from Java
DUBO0205 skull *Ursus torquatus*
DUBO0206 skull *Ursus torquatus*
DUBO0207 skull *Ursus torquatus*
DUBO0208 jawfragment Kedung Brubus spec. Dubois
DUBO0209 premolar
DUBO0210 jawfragment Kedung Brubus spec. Dubois
DUBO0211 jawfragment Kedung Brubus spec. Dubois
DUBO0212 teeth
DUBO0213 tooth
DUBO0214 jawfragment Kedung Brubus spec. Dubois, EDp-1924f
DUBO0215 jawfragment Kedung Brubus spec. Dubois, EDp-1924f
DUBO0216 jawfragment Kedung Brubus spec. Dubois, EDp-1924f
DUBO0217 jawfragment Kedung Brubus spec. Dubois, EDp-1924f
DUBO0218 premolar Dubois spec.
DUBO0219 skull crocodile
DUBO0220 skull crocodile
DUBO0221 skull crocodile
DUBO0222 skull *gavialis*
DUBO0223 skin crocodile
DUBO0224 skin crocodile

DUBO0225 skull *gavialis*
DUBO0226 skin crocodile
DUBO0227 scale patterns of snakes
DUBO0228 skull *gavialis*
DUBO0229 skull *gavialis*
DUBO0230 skull crocodile
DUBO0231 skull crocodile
DUBO0232 skull crocodile
DUBO0233 skull crocodile
DUBO0234 skull *Hylobates*
DUBO0235 jawfragment Kedung Brubus spec. Dubois
DUBO0236 jawfragment Kedung Brubus spec. Dubois, EDp-1924f
DUBO0237 jawfragment Kedung Brubus spec. Dubois, EDp-1924f
DUBO0238 skull *Hylobates*
DUBO0239 skull *Hylobates*
DUBO0240 skull *Hylobates agilis* ♂ +½ natural size, small lens, André de la Pate
DUBO0241 skull *Hylobates agilis,* ♂ ½ natural size, Steinheils Anti-planet 64mm
Diaphr. 5mm
DUBO0242 endocast, EDp-1940b
DUBO0243 endocast, EDp-1940b
DUBO0244 endocast, EDp-1940b
DUBO0245 endocast, EDp-1940b
DUBO0246 femur ??? compare DUBO0018
DUBO0247 femur ???
DUBO0248 femur
DUBO0249 Femur
DUBO0250 femur???
DUBO0251 crocodile vertebrae
DUBO0252 molar of small carnivore
DUBO0253 molar of small carnivore
DUBO0254 fish jaws
DUBO0255 fish jaws
DUBO0256 fish jaws
DUBO0257 crocodile vertebra
DUBO0258 fish jaws
DUBO0259 molar of small carnivore
DUBO0260 molars bear
DUBO0261 bone with crocodile bite
DUBO0262 molar of small carnivore
DUBO0263 crocodile vertebra
DUBO0264 molar of small carnivore
DUBO0265 jaw crocodile
DUBO0266 fossil extinct haring
DUBO0267 schematic cross-cut viper lung
DUBO0268 world map
DUBO0269 rat *faberi* a. skull
DUBO0270 rat *fabe*ri a. skull
DUBO0271 rat *faberi* a. lower jaw
DUBO0272 skull and lower jaw crocodile
DUBO0273 *Acanthion brachyurus* LINNAEUS 1758
DUBO0274 schematic cross-cut viper lung
DUBO0275 snout crocodile
DUBO0276 snout crocodile
DUBO0277 world map upwelling water
DUBO0278 basic lung types
**DUBO0279** *Acanthion brachyurus* LINNAEUS 1758 (page 50)
DUBO0280 3 lizards
DUBO0281 lizard head and foot
DUBO282 plate 9: Trinil site
DUBO0283 skull cap *Pithecanthropus* vs Sangiran.
DUBO0284 skull Sangiran
DUBO0285 femora *Pithecanthropus*
DUBO0286 skull humanoid
DUBO0287 plate 2: Kedung Brubus
DUBO0288 skull *Hippopotamus*
DUBO0289 skull *Hippopotamus*
DUBO0290 skull *Hippopotamus*
DUBO0291 skull *Hippopotamus*
DUBO0292 skull *Hippopotamus*
DUBO0293 skull *Hippopotamus*
DUBO0294 lower jaw ape
DUBO0295 lower jaw ape
DUBO0296 lower jaw ape
DUBO0297 premolar
DUBO0298 list of skeletal parts from Trinil Beds
DUBO0299 *Pithecanthropus* vs sangiran
DUBO0300 drawing brain
DUBO0301 lower jaw ape
**DUBO0302** skull *Hylobates agilis*(page 106)
DUBO0303 skull *Hylobates agilis*
DUBO0304 femur (UV light?)
DUBO0305 femur *Pithecanthropus* Dubois spec.
DUBO0306 femur *Pithecanthropus* Dubois spec.
DUBO0307 femur *Pithecanthropus* Dubois spec.
DUBO0308 femur *Pithecanthropus* Dubois spec.
DUBO0309 femur (glass broken)
DUBO0310 femur (glass broken)
DUBO0311 femur
DUBO0312 femur
DUBO0313 femur *Pithecanthropus* Dubois spec.

| | | |
|---|---|---|
| DUBO0314 | | femur *Pithecanthropus* Dubois spec. |
| DUBO0315 | | molar *Pithecanthropus* Dubois spec. |
| DUBO0316 | | molar *Pithecanthropus* Dubois spec., EDp-1924f |
| DUBO0317 | | molar *Pithecanthropus* Dubois spec., EDp-1924f |
| **DUBO0318** | | Molar *Pithecanthropus* Dubois spec., EDp-1924f (see page 12) |
| DUBO0319 | | premolar *Pithecanthropus* Dubois spec. |
| DUBO0320 | | premolars |
| DUBO0321 | | molar roots *Pithecanthropus* Dubois specs, EDp-1924f |
| DUBO0322 | | jaw fragment Kedung Brubus |
| DUBO0323 | | premolar |
| DUBO0324 | | wortels molars *Pithecanthropus* Dubois specs, EDp-1924f |
| **DUBO0325** | | molars *Pithecanthropus* Dubois specs (page 107) |
| DUBO0326 | | premolars |
| DUBO0327 | | molars and premolars *Pithecanthropus* Dubois specs, EDp-1924f |
| DUBO0328 | | molars and premolars *Pithecanthropus* Dubois specs, EDp-1924f |
| DUBO0329 | | molar roots *Pithecanthropus* Dubois specs, EDp-1924f |
| DUBO0330 | | Molars |
| DUBO0331 | | molars *Pithecanthropus* Dubois specimns |
| DUBO0332 | | Premolars |
| DUBO0333 | | Premolars |
| DUBO0334 | | Premolars |
| DUBO0335 | | 2 bear skulls |
| DUBO0336 | | 2 bear skulls |
| DUBO0337 | | 2 bear skulls |
| DUBO0338 | | 2 bear skulls |
| DUBO0339 | | 2 bear skulls |
| DUBO0340 | ⚙ | drawing hersenen fig. uit boek |
| DUBO0341 | ⚙ | evolution according to Dubois, |
| DUBO0342 | ⚙ | endocasts |
| DUBO0343 | ⚙ | skulls carnivores |
| DUBO0344 | ⚙ | skulls carnivores |
| DUBO0345 | | femur juvenile???? |
| DUBO0346 | | femur juvenile???? |
| DUBO0347 | | femur juvenile???? |
| DUBO0348 | | femur juvenile???? |
| DUBO0349 | | skull crocodile |
| DUBO0350 | | vertebra crocodile |
| DUBO0351 | | lower jaw van tapir |
| DUBO0352 | | crocodile |
| DUBO0353 | | jawfragment *Hippopotamus* |
| DUBO0354 | | lower jaw *Prionailurus bengalensis* |
| DUBO0355 | | fragment fossil? |
| DUBO0356 | | fragment fossil? |
| DUBO0357 | | crocodile |
| DUBO0358 | | carnivore canines |
| DUBO0359 | | crocodile |
| DUBO0360 | | crocodile |
| DUBO0361 | | crocodile |
| DUBO0362 | | crocodile |
| DUBO0363 | | cross-cut (of cast of THE) femur |
| DUBO0364 | | cross-cut (of cast of THE) femur |
| DUBO0365 | | drawing of brain on paper between glass |
| DUBO0366 | | drawing of brain on paper between glass |
| DUBO0367 | X | lower jaw |
| DUBO0368 | | premolar |
| DUBO0369 | | premolar |
| DUBO0370 | | jawfragment Kedung Brubus spec. Dubois |
| DUBO0371 | | surface femur Trinil |
| DUBO0372 | | surface femur Trinil |
| DUBO0373 | | surface femur Trinil |
| DUBO0374 | | surface femur Trinil |
| DUBO0375 | | jaw fragment Kedung Brubus |
| DUBO0376 | | jaw fragment Kedung Brubus |
| DUBO0377 | | jaw fragment Kedung Brubus |
| DUBO0378 | | molars *Pithecanthropus* spec. Dubois |
| DUBO0379 | | jawfragment Kedung Brubus spec. Dubois, EDp-1924f |
| DUBO0380 | | jawfragment Kedung Brubus spec. Dubois, EDp-1924f |
| DUBO0381 | | molars, EDp-1924f |
| DUBO0382 | | jawfragment Kedung Brubus spec. Dubois |
| DUBO0383 | | jawfragment Kedung Brubus spec. Dubois |
| DUBO0384 | | jawfragment Kedung Brubus spec. Dubois |
| DUBO0385 | | jawfragment Kedung Brubus spec. Dubois |
| DUBO0386 | | molars ? |
| DUBO0387 | | molars? |
| DUBO0388 | | molars *Pithecanthropus* spec. Dubois |
| DUBO0389 | | jawfragment Kedung Brubus spec. Dubois, EDp-1924f |
| DUBO0390 | | jawfragment Kedung Brubus spec. Dubois |

| | | |
|---|---|---|
| DUBO0391 | | molar *Pithecanthropus* spec. Dubois |
| DUBO0392 | | molars *Pithecanthropus* spec. Dubois |
| **DUBO0393** | | Anna on boat (page 34) |
| DUBO0394 | | endocast, EDp-1940b |
| DUBO0395 | | endocast, EDp-1924f |
| DUBO0396 | | surface femur recent |
| DUBO0397 | | surface femur recent |
| DUBO0398 | ⊙ | crocodile |
| DUBO0399 | | crocodile |
| DUBO0400 | | crocodile |
| DUBO0401 | | jaw hyena? |
| DUBO0402 | | jaws carnivores |
| DUBO0403 | ⊙ | skull 285 tapir + drawing |
| DUBO0404 | ⊙ | deer skulls |
| DUBO0405 | ⊙ | endocast |
| DUBO0406 | ⊙ | endocast |
| DUBO0407 | | endocast |
| DUBO0408 | | skull crocodile |
| DUBO0409 | | skull crocodile |
| DUBO0410 | | molar tapir |
| DUBO0411 | | fish scale? |
| DUBO0412 | | skull crocodile |
| DUBO0413 | | jaw crocodile |
| DUBO0414 | | jaw crocodile |
| DUBO0415 | | skull crocodile |
| DUBO0416 | | jaw crocodile |
| DUBO0417 | | fish scale? |
| DUBO0418 | | skull crocodile |
| DUBO0419 | | skin crocodile |
| DUBO0420 | | skin crocodile |
| DUBO0421 | | horncore 11201 Bibos |
| DUBO0422 | | horncore 11201 Bibos |
| DUBO0423 | | horncore 11201 Bibos |
| DUBO0424 | | 2 lower jaws tiger recent |
| DUBO0425 | | lower jaw tiger recent |
| DUBO0426 | | twee skulls tiger recent |
| DUBO0427 | | ???? |
| DUBO0428 | | ???? |
| DUBO0429 | | ???? |
| DUBO0430 | ⊙ | orang utan molars |
| DUBO0431 | ⊙ | orang utan molars |
| DUBO0432 | | premolar Dubois spec. hyaena |
| DUBO0433 | | jaw hyaena fossil |
| DUBO0434 | | premolars hyeaena fossil |
| DUBO0435 | | premolar Dubois spec. hyaena |
| DUBO0436 | | premolar Dubois spec. hyaena |
| DUBO0437 | | premolar Dubois spec. hyaena |
| DUBO0438 | | premolar Dubois spec. hyaena |
| DUBO0439 | | premolar Dubois spec. hyaena |
| DUBO0440 | | *Stegodon* molar??/premolar Dubois spec. hyaena |
| DUBO0441 | | premolar Dubois spec. hyaena |
| DUBO0442 | | premolar Dubois spec. hyaena |
| DUBO0443 | | lower jaw hyeana fossil |
| DUBO0444 | | 2 premolars hyaena fossil |
| DUBO0445 | | incisors? |
| DUBO0446 | | incisors *Hippopotamus* recent |
| DUBO0447 | ⊙ | drawing jaw?? |
| DUBO0448 | | incisors *Hippopotamus* recent |
| DUBO0449 | ⊙ | *Hippopotamus* teeth and molars |
| DUBO0450 | | fossil rhino molar |
| DUBO0451 | | fossil rhino molar |
| DUBO0452 | | fossil rhino molar |
| DUBO0453 | | fossil rhino molar |
| **DUBO0454** | | skull horse outside home (see page 104) |
| DUBO0455 | | skull horse outside home |
| **DUBO0456** | | skull horse outside home (see page 104) |
| DUBO0457 | ⊙ | song on *Pithecanthropus* |
| DUBO0458 | ⊙ | song on *Pithecanthropus* |
| DUBO0459 | ⊙ | song on *Pithecanthropus* |
| DUBO0460 | ⊙ | song on *Pithecanthropus* |
| DUBO0461 | ⊙ | Eugène Dubois (cut from photograph Cambridge congres) |
| DUBO0462 | | Kriele (damaged negative) |
| DUBO0463 | ⊙ | skull Ngandong |
| DUBO0464 | ⊙ | skull Ngandong |
| DUBO0465 | | recent femur with exostocis |
| DUBO0466 | ⊙ | skull Ngandong |
| DUBO0467 | | wadjak 2 |
| DUBO0468 | | wadjak 2 |
| DUBO0469 | | jaw Wadjak, spec. Dubois |
| DUBO0470 | | jaw Wadjak, spec. Dubois |
| DUBO0471 | | jaw Wadjak, spec. Dubois |
| DUBO0472 | ⊙ | drawing brain on paper between glass |
| DUBO0473 | | recent femur with exostocis |
| DUBO0474 | | molar *Pithecanthropus* Dubois spec., EDp-1924f |
| DUBO0475 | | molar *Pithecanthropus* Dubois spec., EDp-1924f |
| DUBO0476 | | molar *Pithecanthropus* Dubois spec., EDp-1924f |
| DUBO0477 | X | lower jaw |
| DUBO0478 | X | lower jaw/ |
| DUBO0479 | | molar *Pithecanthropus* Dubois spec. |
| DUBO0480 | | molar *Pithecanthropus* Dubois spec. |
| DUBO0481 | | premolar *Pithecanthropus* Dubois spec. |
| DUBO0482 | | premolars |
| DUBO0483 | | molar *Pithecanthropus* Dubois |

| | | |
|---|---|---|
| | | spec. |
| DUBO0484 | | molar *Pithecanthropus* Dubois spec. |
| DUBO0485 | X | lower jaw |
| DUBO0486 | | molar *Pithecanthropus* Dubois spec. |
| DUBO0487 | ⚙ | drawing of ?sun-calendar???? |
| DUBO0488 | | Shells publ. van Benthem Jutting 1937 |
| DUBO0489 | | Shells publ. van Benthem Jutting 1937 |
| DUBO0490 | | Shells publ. van Benthem Jutting 1937 |
| DUBO0491 | | Shells publ. van Benthem Jutting 1937 |
| DUBO0492 | | Shells publ. van Benthem Jutting 1937 |
| DUBO0493 | | Shells publ. van Benthem Jutting 1937 |
| DUBO0494 | | Shells publ. van Benthem Jutting 1937 |
| DUBO0495 | | Shells publ. van Benthem Jutting 1937 |
| DUBO0496 | | Shells publ. van Benthem Jutting 1937 |
| DUBO0497 | | Shells publ. van Benthem Jutting 1937 |
| DUBO0498 | | Shells publ. van Benthem Jutting 1937 |
| DUBO0499 | | Shells publ. van Benthem Jutting 1937 |
| **DUBO0500** | | Shells publ. van Benthem Jutting 1937 (see page 60) |
| DUBO0501 | | Shells publ. van Benthem Jutting 1937 |
| DUBO0502 | | Shells publ. van Benthem Jutting 1937 |
| DUBO0503 | | Shells publ. van Benthem Jutting 1937 |
| **DUBO0504** | | Shells publ. van Benthem Jutting 1937 (see page 59) |
| DUBO0505 | | Shells publ. van Benthem Jutting 1937 |
| DUBO0506 | | Shells publ. van Benthem Jutting 1937 |
| DUBO0507 | | Shells publ. van Benthem Jutting 1937 |
| DUBO0508 | | Shells publ. van Benthem Jutting 1937 |
| DUBO0509 | | Shells publ. van Benthem Jutting 1937 |
| DUBO0510 | | Shells publ. van Benthem Jutting 1937 |
| DUBO0511 | | Shells publ. van Benthem Jutting 1937 |
| DUBO0512 | | skull cap Dubois van binnen gezien, EDp-1924f |
| DUBO0513 | | Molar *Pithecanthropus* Dubois spec. |
| DUBO0514 | ⚙ | tarsius |
| DUBO0515 | ⚙ | captured/killed Gorilla |
| DUBO0516 | | Heberleins fake skull |
| DUBO0517 | | femur *Stegodon*, EDp-1917b |
| DUBO0518 | | Molar *Pithecanthropus* Dubois spec. |
| DUBO0519 | H | molar *Stegodon* (Sulawesi) |
| DUBO0520 | H | molar *Stegodon* (Sulawesi) |
| DUBO0521 | H | molar *Stegodon* (Sulawesi) |
| DUBO0522 | H | molar *Stegodon* (Sulawesi) |
| DUBO0523 | ⚙ | hoektanden *Celebochourus* Sulawesi Hooijer |
| DUBO0524 | ⚙ | pigs kiezen |
| DUBO0525 | H | pig Sulawesi |
| DUBO0526 | H | pig Sulawesi |
| DUBO0527 | H | pig Sulawesi |
| DUBO0528 | H | pig Sulawesi |
| DUBO0529 | | rat Sulawesi |
| DUBO0530 | | rat Sulawesi |
| DUBO0531 | | rat Sulawesi |
| DUBO0532 | | Mimomys |
| DUBO0533 | | Mimomys |
| DUBO0534 | | Mimomys |
| DUBO0535 | | Mimomys |
| DUBO0536 | | Mimomys |
| DUBO0537 | | *Stegodon* molar??/premolar Dubois spec. hyaena/*Stegodon* molar |
| DUBO0538 | | *Stegodon* molar |
| DUBO0539 | | *Stegodon* molar |
| DUBO0540 | | *Stegodon* molar |
| DUBO0541 | | *Stegodon* molar |
| **DUBO0542** | | Grandma Lojenga, Victor, Jean & Eugènie (see page 37) |
| **DUBO0543** | | Victor, WC, Eugènie, WD & Jean/Jean, Eugènie, Victor (see page 39) |
| **DUBO0544** | | Jean, Eugènie, Victor & Grandma Lojenga (see page 37) |
| **DUBO0545** | | Grandma Lojenga & WE (see page 37) |
| **DUBO0546** | | WF, Eugènie, WC, Jean, Victor & WH (see page 38) |
| **DUBO0547** | | Anna? & Victor (see page 38) |
| **DUBO0548** | | girl with dog (see page 32) |
| **DUBO0549** | | CA, Jean, WC, Eugènie (see page 38) |

**DUBO0550** CA, Jean, WC, Eugènie (see page 38)
**DUBO0551** Eugènie & WI (see page 387)
**DUBO0552** Eugènie, CA, Jean, WC, Victor, CB (see page 38)
**DUBO0553** Victor(brother) + his wife Marie Canoy, WA and her husband (see page 39)
**DUBO0554** Trinette? & grandchild (see page 39)
**DUBO0555** double exposed negative city view. See appendix III.
**DUBO0556** Haarlem? See appendix III.
**DUBO0557** girl and garden ornament (see page 45)
**DUBO0558** Man seated in profile (see page 45)
**DUBO0559** Grandma Lojenga, Victor, Jean & Eugenie (see page 36)
**DUBO0560** WG, Eugenie, WC, Victor, Jean (see page 37)
**DUBO0561** see DUBO0553 (see page 37)
**DUBO0562** Current bedding near Kebon (along the Bengawan) District Sapteh Ngawi (see page 72)
DUBO0563 endocast
DUBO0564 endocast
DUBO0565 skull deer
DUBO0566 skull *Procamelus occidentalis*
DUBO0567 brain ?*Procamelus occidentalis*?
DUBO0568 skull ape
DUBO0569 skull recent (*lagopus*?)
DUBO0570 skull recent
DUBO0571 skull recent
DUBO0572 skull recent
DUBO0573 brain size *Pithecanthropus* vs La Chapelle
DUBO0574 skull recent
DUBO0575 cast skull *Moeritherium*
DUBO0576 skull *Equus caballus* (horse) recent
DUBO0577 cast skull *Moeritherium*
DUBO0578 endocast
DUBO0579 gibbon
DUBO0580 skull recent
DUBO0581 skull *Procamelus*
DUBO0582 brain
DUBO0583 skull ape
DUBO0584 skull dog recent
DUBO0585 skull dog recent
DUBO0586 skull recent
DUBO0587 skull recent
DUBO0588 skull *Equus caballus* (horse) recent
DUBO0589 skull recent
DUBO0590 skull seal recent
DUBO0591 skull recent
DUBO0592 gibbon
DUBO0593 La chapelle
DUBO0594 skull chimp
DUBO0595 skeleton of australian
DUBO0596 skull Rhodesia man
DUBO0597 skull cap *Pithecanthropus*
DUBO0598 skull cap *Pithecanthropus*
DUBO0599 skull cap *Pithecanthropus*
DUBO0600 human skull
DUBO0601 brain size comparison La Chapelle, *Pithecanthropus,* hylobates
DUBO0602 arrangement for photo of drawing of Neanderthal skull
DUBO0603 comparison cranium chimpanzee , La Chapelle and French guy…
DUBO0604 endocast
DUBO0605 orang utan
DUBO0606 *Homo sapiens* jaw vs *Hylobates*
DUBO0607 anatomical drawing
DUBO0608 endocast
DUBO0609 endocast
DUBO0610 *Hylobates* brain
DUBO0611 endocast
DUBO0612 endocast
DUBO0613 comparison brain sizes
DUBO0614 comparison brain sizes
DUBO0615 comparison brain sizes
DUBO0616 comparison brain sizes
DUBO0617 comparison brain sizes
DUBO0618 comparison brain sizes
DUBO0619 endocast
DUBO0620 endocast
DUBO0621 endocast
DUBO0622 endocast
DUBO0623 endocast
DUBO0624 endocast
DUBO0625 endocast
DUBO0626 endocast
DUBO0627 comparison brain sizes
DUBO0628 Map Residence Kediri
DUBO0629 Map Residence Kediri
DUBO0630 Map Residence Kediri
DUBO0631 Map Residence Kediri
DUBO0632 Map Residence Kediri
DUBO0633 Map Residence Kediri
DUBO0634 Map Residence Kediri
DUBO0635 skull
DUBO0636 skull Rhodesia man

| | | |
|---|---|---|
| DUBO0637 | ❂ | comparison cranium sizes |
| DUBO0638 | | skull chimpanzee |
| **DUBO0639** | | skull *Bibos* recent. See appendix III. |
| DUBO0640 | | skull *Duboisia* fossil Dubois spec. |
| DUBO0641 | | skull *Duboisia* fossil Dubois spec. |
| DUBO0642 | ❂ | skull primitive man |
| DUBO0643 | ❂ | skull primitive man |
| DUBO0644 | ❂ | skull Neanderthaler |
| DUBO0645 | ❂ | skull Rhodesia man |
| DUBO0646 | ❂ | skull La Chapelle |
| DUBO0647 | ❂ | anatomy orang utan |
| DUBO0648 | ❂ | anatomie chimpanzee |
| DUBO0649 | ❂ | figure brainweight vs bodyweight, EDp-1922f/1923a |
| DUBO0650 | ❂ | figure, EDp-1922f/1923a |
| DUBO0651 | | figure, EDp-1922f/1923a |
| DUBO0652 | | figure, EDp-1922f/1923a |
| DUBO0653 | ❂ | skull La Chapelle, EDp-1922f/1923a |
| DUBO0654 | ❂ | figure |
| DUBO0655 | ❂ | figure |
| DUBO0656 | ❂ | skull ??? |
| DUBO0657 | ❂ | climbing man, EDp-1926c |
| DUBO0658 | | femur *Stegodon*? |
| DUBO0659 | | climbing man, EDp-1926c |
| DUBO0660 | | femur man |
| DUBO0661 | | femur *Stegodon*, EDp-1917b |
| DUBO0662 | | femur *Stegodon*? |
| DUBO0663 | | femur man |
| DUBO0664 | | femur man |
| DUBO0665 | ❂ | climbing man p-Henry F. Osborne |
| DUBO0666 | | femur man |
| DUBO0667 | | femur cast *Pithecanthropus* |
| DUBO0668 | | femur cast *Pithecanthropus* |
| DUBO0669 | | femur recent |
| DUBO0670 | | femur cast *Pithecanthropus* |
| DUBO0671 | | femur cast *Pithecanthropus* |
| DUBO0672 | | femur *Stegodon* |
| DUBO0673 | | femur *Stegodon* |
| **DUBO0674** | | femur *Stegodon*, EDp-1927b (page 108) |
| DUBO0675 | | femur prox |
| DUBO0676 | | femur prox |
| DUBO0677 | | skull cap *Pithecanthropus* |
| DUBO0678 | | skull cap *Pithecanthropus*, EDp-1924f |
| DUBO0679 | | skull cap *Pithecanthropus*, EDp-1924f |
| DUBO0680 | | skull cap *Pithecanthropus*, EDp-1924f |
| DUBO0681 | | skull cap *Pithecanthropus*, EDp-1924f |
| DUBO0682 | | skull cap *Pithecanthropus*, EDp-1924f |
| DUBO0683 | | skull cap *Pithecanthropus*, EDp-1924f |
| DUBO0684 | | skull cap *Pithecanthropus*, EDp-1924f |
| DUBO0685 | | skull cap *Pithecanthropus*, EDp-1924f |
| DUBO0686 | | skull cap *Pithecanthropus*, EDp-1924f |
| DUBO0687 | | skull cap *Pithecanthropus*, EDp-1924f |
| DUBO0688 | | skull cap *Pithecanthropus*, EDp-1924f |
| DUBO0689 | | skull cap *Pithecanthropus*, EDp-1924f |
| **DUBO0690** | | Trinil plate 8 (page 75) |
| DUBO0691 | | endocast |
| DUBO0692 | | endocast |
| DUBO0693 | | femur *Pithecanthropus* |
| DUBO0694 | | femur *Pithecanthropus* |
| DUBO0695 | X | *Pithecanthropus*, EDp-1926m |
| DUBO0696 | X | *Pithecanthropus*, EDp-1926m |
| DUBO0697 | ❂ | inside skull (von Koenigswald?) |
| DUBO0698 | ❂ | skull sangiran in pieces |
| DUBO0699 | | tortoise fossil *Hardella isoclina* |
| DUBO0700 | X | humerus |
| DUBO0701 | ❂ | skulls and brains |
| DUBO0702 | ❂ | skull *Cynohyenodon cayluxi* |
| DUBO0703 | | Molars and premolar *Pithecanthropus*, specs Dubois |
| DUBO0704 | | skull cap *Pithecanthropus*, EDp-1924f |
| DUBO0705 | | endocast |
| **DUBO0706** | | endocast (see page 107) |
| **DUBO0707** | | Tufwall at Kedung Brubus (see page 65) |
| DUBO0708 | | skull cap *Pithecanthropus*, EDp-1924f |
| **DUBO0709** | | along Kali Ngeto (see page 69) |
| **DUBO0710** | | East of Kedung Brubus (see page 65) |
| **DUBO**0711 | | Detail of DUBO0709? |
| DUBO0712 | | *Homo mojokertensis* VON KOENIGSWALD 1936 |
| DUBO0713 | | *Homo mojokertensis* VON KOENIGSWALD 1936 |
| DUBO0714 | | endocast |
| DUBO0715 | | endocast |
| DUBO0716 | | endocast |

|  | |  |
|---|---|---|
| DUBO0717 | | endocast |
| DUBO0718 | | molar *Elephas hysudrindicus* Dubois spec. |
| DUBO0719 | | molar *Elephas hysudrindicus* Dubois spec. |
| DUBO0720 | | molar *Elephas hysudrindicus* Dubois spec. |
| DUBO0721 | | molar *Elephas hysudrindicus* Dubois spec. |
| DUBO0722 | | molar *Elephas hysudrindicus* Dubois spec. |
| DUBO0723 | | molar *Elephas hysudrindicus* Dubois spec. |
| DUBO0724 | | skull cap *Pithecanthropus,* EDp-1924f |
| DUBO0725 | | skull cap *Pithecanthropus,* EDp-1924f |
| DUBO0726 | | skull cap *Pithecanthropus* |
| DUBO0727 | | skull cap *Pithecanthropus,* EDp-1924f |
| **DUBO0728** | | skull cap *Pithecanthropus,* EDp-1924f (see page 107) |
| DUBO0729 | | skull cap *Pithecanthropus* |
| DUBO0730 | | Molars and premolar *Pithecanthropus,* specs Dubois, EDp-1924f |
| DUBO0731 | ⊙ | brain |
| DUBO0732 | | endocast |
| DUBO0733 | | anatomic drawing *Hippopotamus* skull |
| DUBO0734 | | endocast |
| DUBO0735 | ⊙ | brain medial view |
| DUBO0736 | ⊙ | medial cut skull *Hippopotamus* |
| DUBO0737 | ⊙ | brain |
| DUBO0738 | ⊙ | brain |
| DUBO0739 | ⊙ | brain *Hippopotamus* |
| DUBO0740 | | endocast |
| DUBO0741 | | endocast |
| DUBO0742 | | endocast |
| DUBO0743 | X | femur |
| **DUBO0744** | | stratigraphy sandpit Maarn (see page 114) |
| DUBO0745 | | stratigraphy sandpit Maarn |
| DUBO0746 | | stratigraphy sandpit Maarn |
| **DUBO0747** | | Canoy Herfkens clay quarry (see page 111)Tegelen |
| DUBO0748 | | Stratigraphy at quarry Canoy-Herfkens Tegelen |
| **DUBO0749** | | stratigraphy at quarry Canoy-Herfkens Tegelen (see page 113) |
| DUBO0750 | | stratigraphy at quarry Canoy-Herfkens Tegelen |
| **DUBO0751** | | quarry Maalbeek east of Belfeld (see page 113) |
| DUBO0752 | | *Hypolohus sephen* 2475 |
| DUBO0753 | | *Hypolohus sephen* 2475 |
| DUBO0754 | | *Trygon polylepis 2470* |
| DUBO0755 | | *Hypolohus sephen* 2475 |
| DUBO0756 | | *Hypolohus sephen* 2474 |
| DUBO0757 | | *Hypolohus?* |
| DUBO0758 | | skull cap *Pithecanthropus,* EDp-1924f |
| DUBO0759 | | skull cap *Pithecanthropus,* EDp-1924f |
| DUBO0760 | | skull cap *Pithecanthropus,* EDp-1924f |
| DUBO0761 | | skull cap *Pithecanthropus* |
| **DUBO0762** | | skull cap *Pithecanthropus* (see page 106) |
| DUBO0763 | | skull cap *Pithecanthropus* |
| DUBO0764 | | femur *Pithecanthropus* |
| DUBO0765 | | femur *Pithecanthropus* |
| DUBO0766 | | skeleton *Hippopotamus* recent |
| DUBO0767 | | molar *Elephas hysudrindicus* Dubois spec. |
| DUBO0768 | | crocodile? |
| DUBO0769 | | canines *Celebochoerus hekereeni,* Sulawesi |
| DUBO0770 | | jaws *Celebochoerus* ? |
| DUBO0771 | H | *Nothosaurus* Winterswijk |
| DUBO0772 | H | humerus *Nothosaurus* Winterswijk |
| DUBO0773 | H | lower jaw *Nothosaurus* Winterswijk |
| **DUBO0774** | ⊙ H | Stratigraphy Foulkes Gravelpit Kempston. Photo van Heekeren. See appendix III |
| DUBO0775 | | femur Wajak, marble quarry near Wajak, Java |
| DUBO0776 | | femur Wajak, marble quarry near Wajak, Java |
| **DUBO0777** | ⊙ | Henry Fairfield Osborne (see page 123) |
| DUBO0778 | ⊙ | reconstruction fauna Tegelen |
| DUBO0779 | | Winckler |
| DUBO0780 | | 2 pieces of bone… |
| DUBO0781 | | tortoise *Hardella isoclina* Dubois spec. nr. 2722 |
| DUBO0782 | | schildpad *Hardella isoclina* Dubois spec. nr. 2722 |
| DUBO0783 | | schildpad *Hardella isoclina* Dubois spec. nr. 2722 |
| DUBO0784 | | tortoise recent (stereophoto) |
| DUBO0785 | | endocast |
| DUBO0786 | | endocast |
| DUBO0787 | | endocast |

| | |
|---|---|
| **DUBO0788** | endocast (see page 107) |
| DUBO0789 | endocast |
| DUBO0790 | endocast |
| DUBO0791 | endocast |
| DUBO0792 | endocast |
| DUBO0793 | endocast |
| DUBO0794 | endocasts |
| DUBO0795 | endocast |
| DUBO0796 | endocast |
| DUBO0797 | Siwalik 1894 expedition camels |
| **DUBO0798** | Siwalik 1894 expedition group (see page 22) |
| **DUBO0799** | Siwalik 1894 expedition elephant (see page 19) |
| **DUBO0800** | Siwalik 1894 expedition Ram Phal (see page 21) |
| **DUBO0801** | Siwalik 1894 expedition girl (see page 25) |
| DUBO0802 | endocast |
| DUBO0803 | *Phenocodus primaevus* |
| DUBO0804 | anatomic preparate |
| DUBO0805 | endocast |
| DUBO0806 | table on taxonomic order |
| DUBO0807 | femora *Pithecanthropus* Dubois |
| DUBO0808 | map of Eocene Europe |
| DUBO0809 | map of Oligocene Europe |
| DUBO0810 | world map middle devoon |
| DUBO0811 | map of Pliocene Europe |
| DUBO0812 | map of Pliocene Europe |
| DUBO0813 | map of Pliocene Europe |
| DUBO0814 | wereld kaart midden devoon |
| DUBO0815 | map of Oligocene Europe |
| DUBO0816 | world map Triassic |
| DUBO0817 | world map Cretaceous |
| DUBO0818 | map of Plioceen Europe |
| DUBO0819 | map of Mioceen Europe |
| DUBO0820 | upper jaw *Hippopotamus* fossil |
| DUBO0821 | skulls fossil Dubois |
| **DUBO0822** | fossil *Hippopotamus* jaws (see page 57) |
| **DUBO0823** | upper jaw fossil rhinoceros (see page 50) |
| DUBO0824 | fossil *Hippopotamus* |
| DUBO0825 | fossil upper jaw rhinoceros |
| DUBO0826 | rhinoceros skulls recent |
| **DUBO0827** | fossil *Hippopotamus* lower jaws (see page 120) |
| DUBO0828 | fossil *Hippopotamus* skull Dubois Kedung Brubus |
| DUBO0829 | skulls fossil Dubois collection |
| **DUBO0830** | fossil rhinoceros upper jaw (see page 57) |
| DUBO0831 | fossil *Hippopotamus* skull Dubois Kedung Brubus |

| | |
|---|---|
| **DUBO0832** | Siwalik 1894 expedition (see page 23) |
| **DUBO0833** | Siwalik 1894 expedition camels (see page 20) |
| DUBO0834 | Siwalik 1894 expedition camels |
| **DUBO0835** | Siwalik 1894 expedition women (see page 26) |
| **DUBO0836** | Siwalik 1894 expedition women (see page 26) |
| DUBO0837 | Siwalik 1894 expedition camel |
| **DUBO0838** | Siwalik 1894 expedition group (see page 23) |
| DUBO0839 | Siwalik 1894 expedition women |
| **DUBO0840** | Siwalik 1894 expedition women (see page 25) |
| DUBO0841 | Siwalik 1894 expedition group |
| **DUBO0842** | Siwalik 1894 expedition men (see page 27) |
| **DUBO0843** | Siwalik 1894 expedition team (see page 21) |
| DUBO0844 | world map Cambrium |
| DUBO0845 | world map middle Jurassic |
| DUBO0846 | world map upper Devone |
| DUBO0847 | world map Portland period |
| DUBO0848 | world map lower Carbone |
| DUBO0849 | world map middle Eocene |
| DUBO0850 | world map middle Silure |
| DUBO0851 | world map lower Miocene |
| DUBO0852 | world map middle Carbone |
| DUBO0853 | world map lower Jurassic |
| DUBO0854 | world map upper Jurrasic |
| DUBO0855 | world map lower Triassic |
| DUBO0856 | world map ? middle Devone? |
| DUBO0857 | world map upper Silure |
| DUBO0858 | map of Oligocene Europe |
| DUBO0859 | antropological death mask Melanesian male circa 20 y |
| DUBO0860 | antropological death mask |
| DUBO0861 | antropological death mask Micronesian female circa 18 y |
| DUBO0862 | cover van Recherche ossement fossiles |
| DUBO0863 | antropological death mask Micronesian female circa 38 y |
| DUBO0864 | antropological death mask Micronesian female circa 18 y |
| DUBO0865 | antropological death mask Micronesian female circa 38 y |
| DUBO0866 | antropological death mask Melanesian male circa 20 y |
| DUBO0867 | antropological death mask |
| DUBO0868 | orang utan molars |

| | |
|---|---|
| DUBO0869 | antlers *Axis lydekkeri* |
| DUBO0870 | skulls *Duboisia* |
| DUBO0871 | skull *Bibos palaeosondaicus* |
| DUBO0872 | skull *Bibos palaeosondaicus* |
| DUBO0873 | endocast |
| DUBO0874 | endocast |
| DUBO0875 | Shell publ. van Benthem Jutting 1937 |
| DUBO0876 | endocast |
| DUBO0877 | endocast |
| DUBO0878 | endocast |
| DUBO0879 | femur *Pithecanthropus* |
| DUBO0880 | femur *Pithecanthropus* |
| DUBO0881 | femur *Pithecanthropus* |
| **DUBO0882** | femur *Pithecanthropus* (see page 106) |
| DUBO0883 | femur *Pithecanthropus* |
| DUBO0884 | femur *Pithecanthropus* |
| DUBO0885 | femur *Pithecanthropus* |
| DUBO0886 | femur *Pithecanthropus* |
| DUBO0887 | femur *Pithecanthropus* |
| DUBO0888 | femur *Pithecanthropus* |
| DUBO0889 | femur *Pithecanthropus* |
| DUBO0890 | femur *Pithecanthropus* |
| **DUBO0891** | Gerardine and her kids? (see page 39) |
| **DUBO0892** | boyservant (see page 15) |
| **DUBO0893** | three kids (see page 42) |
| **DUBO0894** | Daughters of Gerardine? (see page 41) |
| **DUBO0895** | Anna Grace Boyd?? (see page 18) |
| **DUBO0896** | Djongas Nassi (see page 16) |
| **DUBO0897** | Manservant (see page 15) |
| **DUBO0898** | garden Netherl. (see page 44) |
| **DUBO0899** | streetscene ± 1900 (see appendix III) |
| **DUBO0900** | Prentice (see page 17) |
| **DUBO0901** | Boyd (see page 17) |
| DUBO0902 | Dubois, excerpt from Cambridge, Smith and Keith |
| DUBO0903 | Dubois peinting at the age of 70 |
| DUBO0904 | De Winter |
| DUBO0905 | houses in Den Haag, Duinoord +- 1900 |
| **DUBO0906** | De Winter (see page 11) |
| **DUBO0907** | veranda with fossils (see page 15) |
| DUBO0908 | Kriele |
| **DUBO0909** | veranda with fossils (see page 96) |
| **DUBO0910** | veranda with fossils (see page 14) |
| **DUBO0911** | Kriele (see page 11) |
| DUBO0912 | gorilla skull |
| DUBO0913 | twee skulls ape |
| DUBO0914 | *Anthropopithecus troglodytus* |
| DUBO0915 | *Anthropopithecus troglodytus* |
| DUBO0916 | *Anthropopithecus troglodytus* |
| DUBO0917 | twee skulls monkeys |
| DUBO0918 | skull Papoea |
| DUBO0919 | *Anthropopithecus troglodytus* |
| DUBO0920 | bovine upper jaws |
| DUBO0921 | skull *Hippopotamus* fossil Kedung Brubus |
| DUBO0922 | fragmenten skull Bibos |
| **DUBO0923** | jaws several species (see page 54) |
| DUBO0924 | lower jaw *Hyaena* |
| **DUBO0925** | lower jaw *Hyaena*(see page 55) |
| DUBO0926 | lower jaw *Hyaena* |
| DUBO0927 | lower jaw *Hyaena* |
| **DUBO0928** | lower jaw *Panthera tigris* (see page 50) |
| **DUBO0929** | fossil tapir lower jaw molars (see page 57) |
| DUBO0930 | fossil *Gavialis* Trinil Dubois specs |
| DUBO0931 | fossil *Gavialis* Trinil Dubois specs |
| DUBO0932 | *Crocodilus* Trinil |
| DUBO0933 | *Crocodilus* Trinil |
| DUBO0934 | *Crocodilus* 3978 |
| DUBO0935 | *Crocodilus* |
| **DUBO0936** | jaws *Stegodon*(see page 51) |
| **DUBO0937** | tanden *Stegodon* (see page 51) |
| **DUBO0938** | lower jaws *Duboisia* and *Axis* (see page 53) |
| DUBO0939 | lower jaws *Duboisia* and *Axis* |
| **DUBO0940** | upper jaws *Duboisia* and *Lutra* (see page 56) |
| DUBO0941 | skull *Stegodon* Kedung Panas |
| DUBO0942 | skull *Stegodon* Trinil |
| DUBO0943 | lower jaws *Stegodon* |
| DUBO0944 | lower jaws *Stegodon* |
| **DUBO0945** | teeth *Stegodon* (see page 56) |
| DUBO0946 | femur *Pithecanthropus* |
| DUBO0947 | femur *Pithecanthropus* |
| DUBO0948 | cross-cut femur? |
| DUBO0949 | femur *Pithecanthropus* |
| DUBO0950 | skull *Hylobates*? |
| DUBO0951 | lower jaw monkey |
| DUBO0952 | skull ape |
| DUBO0953 | skull ape |
| DUBO0954 | femur *Pithecanthropus* |
| DUBO0955 | skull cap *Pithecanthropus* spec. Dubois in foto opstelling |
| **DUBO0956** | skull cap *Pithecanthropus* (see page 105) |
| DUBO0957 | skull cap *Pithecanthropus* |

| | |
|---|---|
| **DUBO0958** | footbones *Manis* (see page 55) |
| **DUBO0959** | feet *Manis* (see page 55) |
| DUBO0960 | footbones *Manis* |
| DUBO0961 | footbones *Manis* |
| DUBO0962 | skeleton *Manis temminckii* |
| DUBO0963 | footbones *Manis* |
| DUBO0964 | skeleton *Manis aurita* |
| DUBO0965 | skeleton *Manis aurita* |
| DUBO0966 | skeleton *Manis* |
| DUBO0967 | skeleton *Manis temminckii* |
| DUBO0968 | skeleton *Manis longicaudata* |
| DUBO0969 | skull *Manis gigantea* |
| **DUBO0970** | skull *Epileptobos* (see page 121) |
| DUBO0971 | kaken *Hippopotamus* |
| DUBO0972 | skull *Axis lydekkeri* |
| DUBO0973 | skull *Epileptobos* |
| DUBO0974 | skulls *Duboisia* |
| DUBO0975 | skulls *Bibos* |
| **DUBO0976** | skulls *Bubalus palaeokerabau* (see page 54) |
| **DUBO0977** | antlers *Axis lydekkeri* (see page 52) |
| **DUBO0978** | skulls *Bibos* (see page 53) |
| DUBO0979 | skulls bovines |
| DUBO0980 | skulls *Bubalus palaeokerabau* |
| **DUBO0981** | skulls *Duboisia* (see page 52) |
| DUBO0982 | Willemien Lojenga, Jean, Anna Lojenga |
| **DUBO0983** | Anna Lojenga in garden (see page 42) |
| **DUBO0984** | Eugènie (see page 42) |
| **DUBO0985** | Eugènie (see page 40) |
| DUBO0986 | Willemien Lojenga, CC, Victor, Anna Lojenga, Jean |
| **DUBO0987** | Willemien Lojenga, Victor, Anna Lojenga (see page 41) |
| **DUBO0988** | Mere Marie-Angelique (see page 40) |
| **DUBO0989** | Willemien Lojenga, CC, Victor, Anna Lojenga, Jean (see page 41) |
| **DUBO0990** | Janet Boyd?? (see page 41) |
| DUBO0991 | Eugènie |
| **DUBO0992** | Eugènie (see page 40) |
| **DUBO0993** | Anna Lojenga (see page 39) |
| DUBO0994 | maxilla Wadjak 2 |
| DUBO0995 | Skull wadjak 1 |
| DUBO0996 | maxilla Wadjak 2 |
| DUBO0997 | Skull wadjak 1 |
| DUBO0998 | maxilla Wadjak 2 |
| DUBO0999 | Skull wadjak 1 |
| DUBO1000 | Skull wadjak 1 |
| DUBO1001 | cut through cast? |
| DUBO1002 | cut through cast? |
| DUBO1003 | cut through cast? |
| DUBO1004 | femur distal |
| DUBO1005 | femur distal |
| DUBO1006 | femur proximal |
| DUBO1007 | femur proximal |
| DUBO1008 | femur and skeleton man |
| DUBO1009 | cut through cast? |
| DUBO1010 | endocast |
| DUBO1011 | endocast |
| DUBO1012 | endocast |
| DUBO1013 | endocast |
| DUBO1014 | endocast |
| DUBO1015 | endocast |
| DUBO1016 | skull cap vs *Hylobates* |
| DUBO1017 | Endocast |
| DUBO1018 | Endocast, EDp-1924f |
| DUBO1019 | endocast, EDp-1924f |
| DUBO1020 | endocast, EDp-1924f |
| DUBO1021 | endocast, EDp-1924f |
| **DUBO1022** | tortoise *Hardella isoclina* (see page 58) |
| **DUBO1023** | seeds Tegelen (see page 112) |
| DUBO1024 | skull *Symia satyrus* |
| DUBO1025 | femur proximal |
| DUBO1026 | tortoise *Hardella isoclina* |
| DUBO1027 | femur proximal |
| DUBO1028 | femur *Pithecanthropus* |
| DUBO1029 | femur proximal |
| DUBO1030 | femur proximal |
| DUBO1031 | stratigraphy Tegelen? |
| DUBO1032 | stratigraphy Tegelen? |
| DUBO1033 | erratic boulder from Moorbeek |
| DUBO1034 | ?stratigraphy Tegelen? |
| DUBO1035 | sand? quarry |
| DUBO1036 | =DUBO1031 stratigraphy Teg |
| DUBO1037 | erratic boulder from Moorbeek |
| DUBO1038 | antler fragments *Eucladoceros* |
| **DUBO1039** | antler of *Eucladoceros* in situ Tegelen (see page 110) |
| DUBO1040 | antler *Eucladoceros* Tegelen |
| DUBO1041 | antler fragment Netherlands |
| DUBO1042 | antler fragment Netherlands |
| DUBO1043 | antlerfragmenten *Cervus tegulensis* |
| DUBO1044 | antler *Cervus tegulensis* |
| DUBO1045 | antler *Cervus tegulensis* |
| DUBO1046 | antler *Cervus tegulensis* |
| DUBO1047 | antler fragment Netherlands |
| DUBO1048 | antler fragment Netherlands |
| DUBO1049 | X femora |
| DUBO1050 | X femur |
| DUBO1051 | X femur, EDp-1926m |
| DUBO1052 | X femur |

| | | |
|---|---|---|
| DUBO1053 | X | femur, EDp-1926m |
| DUBO1054 | X | femur, EDp-1926m |
| DUBO1055 | X | teeth and jawfragment Kedung Brubus |
| DUBO1056 | X | femur, EDp-1926m |
| DUBO1057 | X | femur |
| DUBO1058 | X | teeth and jawfragment Kedung Brubus |
| DUBO1059 | X | femora |
| DUBO1060 | | unknown quarry |
| DUBO1061 | | unknown quarry |
| DUBO1062 | | unknown quarry |
| DUBO1063 | | unknown quarry |
| DUBO1064 | | unknown quarry |
| DUBO1065 | | unknown quarry |
| DUBO1066 | | unknown quarry |
| DUBO1067 | | unknown quarry |
| DUBO1068 | | endocast |
| DUBO1069 | | endocast |
| DUBO1070 | | endocast |
| DUBO1071 | | endocast |
| DUBO1072 | | endocast |
| DUBO1073 | | endocast |
| DUBO1074 | | endocast |
| DUBO1075 | | figure |
| DUBO1076 | | maxilla Wadjak 2 |
| DUBO1077 | | maxilla Wadjak 2 |
| DUBO1078 | | skull man ?? |
| DUBO1079 | | *Homo modjokertensis* VON KOENIGSWALD 1936 |
| DUBO1080 | | *Homo modjokertensis* VON KOENIGSWALD 1936 |
| DUBO1081 | | jaw *Pithecanthropus* VON KOENIGSWALD uit boek |
| DUBO1082 | | quarry |
| DUBO1083 | | Hondsrug |
| **DUBO1084** | | Hondsrug (see page 114) |
| **DUBO1085** | | stratigraphy near Savelsbeek, see appendix III |
| DUBO1086 | | erratic boulder from Moorbeek |
| **DUBO1087** | | erratic boulder from Moorbeek, see appendix III |
| DUBO1088 | ⚙ | reconstruction Java Man |
| DUBO1089 | ⚙ | Manouvrier |
| DUBO1090 | X | femur, EDp-1926m |
| DUBO1091 | | crocodile vertebrae |
| **DUBO1092** | | kaken van een dog (see page 9) |
| DUBO1093 | ⚙ | text on *Manis* |
| DUBO1094 | | femora *Pithecanthropus* |
| DUBO1095 | | femora *Pithecanthropus* |
| DUBO1096 | ⚙ | reconstruction Java Man |
| DUBO1097 | ⚙ | reconstruction Java Man |
| DUBO1098 | | Lower jaw Wadjak 2 spec. Dubois |
| DUBO1099 | | skull wadjak 1 spec. Dubois |
| **DUBO1100** | | Hoekgrot (see page 61) |
| **DUBO1101** | | view on Hoekgrot (see page 61) |
| **DUBO1102** | | view from Hoekgrot (see page 62) |
| **DUBO1103** | | view from Hoekgrot (see page 63) |
| **DUBO1104** | | view from Hoekgrot (see page 63) |
| **DUBO1105** | | view on hoekgrot (see page 61) |
| **DUBO1106** | | view on Wadjak (see page 61) |
| **DUBO1107** | | view on Wadjak (see page 62) |
| **DUBO1108** | | view on Wadjak (see page 62) |
| **DUBO1109** | | view on Hoekgrot (see page 61) |
| DUBO1110 | | endocast |
| DUBO1111 | | endocast |
| DUBO1112 | | endocast |
| DUBO1113 | | endocast |
| DUBO1114 | | endocast |
| DUBO1115 | | endocasts |
| DUBO1116 | | skull cap vs ape |
| DUBO1117 | | two cat skulls |
| DUBO1118 | | two cat skulls |
| DUBO1119 | | *Gavialis* |
| DUBO1120 | | *Gavialis* |
| DUBO1121 | | *Gavialis* skull |
| DUBO1122 | | *Gavialis* skull |
| DUBO1123 | | skull crocodile |
| DUBO1124 | | skull crocodile |
| DUBO1125 | | two cat skulls |
| DUBO1126 | | skull crocodile |
| DUBO1127 | | two cat skulls |
| DUBO1128 | | endocast |
| DUBO1129 | | endocast |
| DUBO1130 | | endocast |
| DUBO1131 | ⚙ | sketch brains *A. troglodytes* |
| DUBO1132 | ⚙ | sketch brains *H. agilis* |
| **DUBO1133** | | stratigraphy near railroad (see page 74) |
| **DUBO1134** | | monument Trinil (see page 77) |
| **DUBO1135** | ⚙ | stratigraphy near railroad (see page 73) |
| **DUBO1136** | ⚙ | stratigraphy near railroad (see page 73) |
| **DUBO1137** | | stratigraphy near railroad (see page 69) |
| DUBO1138 | X | femora |
| DUBO1139 | X | femur |
| DUBO1140 | X | femora |
| DUBO1141 | X | femur |
| DUBO1142 | X | femora |
| DUBO1143 | X | femur |

|  | |  |
|---|---|---|
| DUBO1144 | X | femur |
| DUBO1145 | X | femur |
| DUBO1146 | | femur *Pithecanthropus* |
| DUBO1147 | | femur *Pithecanthropus* |
| DUBO1148 | | femur *Pithecanthropus* |
| DUBO1149 | | femur distal |
| DUBO1150 | | femur distal, EDp-1926m |
| DUBO1151 | | femur *Pithecanthropus* |
| DUBO1152 | | femur *Pithecanthropus* |
| DUBO1153 | | femur *Pithecanthropus* |
| DUBO1154 | ☼ | *Protoceratops andrewsi* |
| DUBO1155 | ☼ | Ngandong skull |
| DUBO1156 | ☼ | skull Wajak |
| DUBO1157 | ☼ | reconstruction Java Man |
| DUBO1158 | | skeleton *Manis* |
| DUBO1159 | | veranda met fossilen (see DUBO0907 ) |
| DUBO1160 | | fossils *Manis* Dubois specs |
| DUBO1161 | | endocast |
| DUBO1162 | | femur |
| DUBO1163 | | skull fragments *Duboisia* |
| DUBO1164 | | endocast |
| DUBO1165 | | endocast |
| DUBO1166 | | femur |
| DUBO1167 | | endocast |
| DUBO1168 | | endocast |
| DUBO1169 | | endocast |
| DUBO1170 | | endocast |
| DUBO1171 | | endocast |
| DUBO1172 | | endocast |
| DUBO1173 | | endocast |
| DUBO1174 | | endocast |
| **DUBO1175** | | endocast (see page 69) |
| DUBO1176 | | femur, photo cranium, skull *Hylobates* (broken negative) |
| DUBO1177 | | antler *Eucladoceros tegulensis* |
| DUBO1178 | | antler *Eucladoceros tegulensis* |
| DUBO1179 | | antler *Eucladoceros tegulensis* |
| DUBO1180 | | antler *Cervus rhenanus* |
| DUBO1181 | | antler fragments *Eucladoceros tegulensis* |
| DUBO1182 | | antler *Eucladoceros tegulensis* |
| DUBO1183 | | lower jaw Wajak 2 |
| DUBO1184 | | lower jaw Wajak 2 |
| DUBO1185 | | lower jaw Wajak 2 |
| DUBO1186 | | Endocast |
| DUBO1187 | | Endocast |
| DUBO1188 | | Endocast |
| DUBO1189 | | lower jaw Wajak 2 |
| DUBO1190 | | femur ??? |
| DUBO1191 | | antler *Cervus rhenanus* |
| DUBO1192 | | lower jaw Wajak 2 |
| DUBO1193 | | skull cap ventral view unprepared |
| DUBO1194 | | skull *Pithecanthropus*, EDp-1924f |
| DUBO1195 | | skull *Pithecanthropus*, EDp-1924f |
| DUBO1196 | | skull *Pithecanthropus*, EDp-1924f |
| DUBO1197 | | skull *Pithecanthropus*, EDp-1924f |
| DUBO1198 | | teeth carnivore |
| DUBO1199 | | teeth carnivore |
| DUBO1200 | | fossil *Gavialis* Trinil no 9? |
| DUBO1201 | | fossil *Gavialis* Trinil no 9 |
| DUBO1202 | | fossil *Gavialis* Trinil no 9 |
| DUBO1203 | | skull fossile crocodile, Trinil, no 12 |
| DUBO1204 | | skull fossile crocodile, Trinil, no 12 |
| DUBO1205 | | skull fossile crocodile, Trinil, no 12 |
| DUBO1206 | | lower jaw crocodile fossil, Trinil, no. 15 |
| DUBO1207 | | skull fossil crocodile, Trinil, no 12 |
| DUBO1208 | | skull fossil *Gavialis* |
| DUBO1209 | | skeleton recent man |
| DUBO1210 | | skeleton ape |
| DUBO1211 | | skeleton ape |
| DUBO1212 | | skeleton ape |
| DUBO1213 | | skeleton ape |
| DUBO1214 | | skull crocodile |
| DUBO1215 | ☼ | crocodile |
| DUBO1216 | | head crocodile |
| DUBO1217 | | skull and lower jaw crocodile |
| DUBO1218 | | crocodile |
| DUBO1219 | | crocodile |
| DUBO1220 | | skull crocodile fossil Trinil |
| DUBO1221 | | skull crocodile fossil Trinil |
| DUBO1222 | | skull crocodile |
| DUBO1223 | ☼ | objects carried |
| DUBO1224 | ☼ | two lizards |
| DUBO1225 | H | bovine molars in sediment |
| DUBO1226 | ☼ | two lizards |
| DUBO1227 | | upper jaw rhinoceros fossil |
| DUBO1228 | | upper jaw rhinoceros fossil |
| DUBO1229 | H | lower jaw sabertooth tiger |
| DUBO1230 | H | lower jaw sabertooth tiger |
| DUBO1231 | H | lower jaw sabertooth tiger |
| DUBO1232 | H | lower jaw *Elephas celebensis* with tooth |
| DUBO1233 | H | lower jaw *Elephas celebensis* with tooth |
| DUBO1234 | H | lower jaw *Elephas celebensis* with tooth |

DUBO1235 H lower jaw *Elephas celebensis* with tooth
DUBO1236 H lower jaw *Elephas celebensis* with tooth
DUBO1237 H lower jaw *Elephas celebensis* with tooth, Tjeleko
DUBO1238 H Fossil
DUBO1239 H Fossil
DUBO1240 H lower jaw *Elephas celebensis*
DUBO1241 H lower jaw *Elephas celebensis*
**DUBO1242** rund Siwaliks spec. Dubois nr 3107 (see page 27)
**DUBO1243** skull *Bibos paleosondaicus* nr. 2812 (see page 120)
DUBO1244 skull *Epileptobos* spec. Dubois
DUBO1245 skull *Epileptobos* spec. Dubois
DUBO1246 skull *Epileptobos* spec. Dubois
DUBO1247 skull *Epileptobos* spec. Dubois
DUBO1248 skull *Epileptobos* spec. Dubois
DUBO1249 skull *Epileptobos* spec. Dubois
DUBO1250 skull *Epileptobos* spec. Dubois
DUBO1251 molars and premolar *Pithecanthropus* specs Dubois
DUBO1252 molars and premolar *Pithecanthropus* specs Dubois
DUBO1253 Kriele and De Winter
**DUBO1254** Boetak (see page 70)
**DUBO1255** Boetak (see page 71)
DUBO1256 Kali Brubus near Kedung Madoh
**DUBO1257** Kali Brubus near Kedung Madoh (see page 67)
DUBO1258 Kali Ngeto near Kedung Brubus
**DUBO1259** Tanah cleft, Bangle (see page 66)
DUBO1260 Kali Ngeto near Kedung Brubus
DUBO1261 unknown stratigraphy, probably near Kedung Brubus
DUBO1262 Tanah cleft, Bangle
**DUBO1263** kali Ngeto near Kedung Ampel (see page 65)
DUBO1264 Kali Ngeto near Kedung Brubus
**DUBO1265** Boetak (see page 71)
**DUBO1266** Boetak (see page 71)
**DUBO1267** Boetak (see page 71)
**DUBO1268** Boetak (see page 71)
**DUBO1269** Kali Ngeto near Kedung Brubus (see page 68)
DUBO1270 Current bedding near Kedung Brubus
DUBO1271 Kali Ngeto near Kedung Brubus
**DUBO1272** Ravine of the Kali Brubus (see page 64)
**DUBO1273** view on the Kendeng (see page 78)
**DUBO1274** Kali Ngeto near Kedung Brubus (see page 68)
DUBO1275 Current bedding near Kedung Brubus
**DUBO1276** Siwalik 1894 expedition (see page 24)
**DUBO1277** Siwalik 1894 expedition (see page 24)
**DUBO1278** Boetak (see page 70)
**DUBO1279** Plate 5 top current bedding Boetak (see page 72)
**DUBO1280** Stereorthoscope (see page 103)
DUBO1281 molars and premolar *Pithecanthropus* specs Dubois
**DUBO1282** Boetak (see page 71)
**DUBO1283** Solo rivir (see page 67)
DUBO1284 Gunung Lawa
DUBO1285 Kali Ngeto near Kedung Brubus
DUBO1286 Current bedding near Kedung Brubus
**DUBO1287** Plate 5 top current bedding Boetak (see page 72)
DUBO1288 Current bedding near Kedung Brubus
**DUBO1289** Tanah cleft, Bangle (see page 65)
DUBO1290 Solo rivir
DUBO1291 Kali Brubus near Kedung Madoh
DUBO1292 killed orang utan
DUBO1293 huge Pithon with prey inside
DUBO1294 skull ape
DUBO1295 skull fossil crocodile spec. Dubois Trinil (broken negative)
DUBO1296 skull fossil *Gavialis* spec. Dubois Trinil
DUBO1297 Skull cap
DUBO1298 skull *Pithecanthropus*
DUBO1299 skull Hylobates
**DUBO1300** Eugènie? (see page 42)
DUBO1301 endocast
DUBO1302 skull ape
**DUBO1303** skull cap unprepared in ventral view (see page 12)
**DUBO1304** garden (see page 44)
DUBO1305 femur *Pithecanthropus* and recente
DUBO1306 femur *Pithecanthropus* and recente
DUBO1307 microcephalic skull ???
**DUBO1308** Tanah cleft, Bangle (see page 65)
DUBO1309 femur *Pithecanthropus* and recent
DUBO1310 fossil fragments ???
DUBO1311 garden see also DUBO1304
**DUBO1312** femur *Pithecanthropus* (see page

| | | |
|---|---|---|
| | | 30) |
| DUBO1313 | | endocast |
| **DUBO1314** | | tree Indonesia (see page 17) |
| DUBO1315 | | treefern |
| DUBO1316 | | skull cap *Pithecanthropus,* EDp-1924f |
| DUBO1317 | | skull cap *Pithecanthropus,* EDp-1924f |
| **DUBO1318** | | Endocast, EDp-1924f (see page 88) |
| DUBO1319 | | skull cap *Pithecanthropus,* EDp-1924f |
| **DUBO1320** | | Endocast, EDp-1924f (see page 89) |
| **DUBO1321** | | Endocast, EDp-1924f (see page 90) |
| **DUBO1322** | | Endocast, EDp-1924f (see page 90) |
| **DUBO1323** | | Endocast, EDp-1924f (see page 88) |
| **DUBO1324** | | microcephalic skull spec. Dubois (see page 87) |
| **DUBO1325** | X | hand Eugènie (see page 43) |
| DUBO1326 | | skull cap *Pithecanthropus,* EDp-1924f |
| DUBO1327 | | femur dist |
| DUBO1328 | | skull cap *Pithecanthropus* |
| DUBO1329 | | skull cap *Pithecanthropus* |
| DUBO1330 | | skull cap *Pithecanthropus* |
| DUBO1331 | | femur dist |
| DUBO1332 | | skull cap *Pithecanthropus* |
| DUBO1333 | | skull cap *Pithecanthropus* |
| DUBO1334 | | skull cap *Pithecanthropus,* EDp-1924f |
| DUBO1335 | | skull cap *Pithecanthropus,* EDp-1924f |
| DUBO1336 | | skull cap *Pithecanthropus,* EDp-1924f |
| DUBO1337 | | skull cap *Pithecanthropus* |
| DUBO1338 | | skull cap *Pithecanthropus* |
| DUBO1339 | | endocast |
| DUBO1340 | | endocast |
| DUBO1341 | | endocast |
| DUBO1342 | ⚙ | endocast sizes |
| DUBO1343 | | endocast |
| DUBO1344 | | endocast, EDp-1924f |
| DUBO1345 | | endocast |
| DUBO1346 | | endocast |
| DUBO1347 | | endocast |
| DUBO1348 | | endocast |
| DUBO1349 | | endocast |
| DUBO1350 | | endocast |
| DUBO1351 | | antlers *Cervus rhenanus* Tegelen |
| DUBO1352 | | endocast |
| DUBO1353 | | endocast |
| DUBO1354 | | endocast |
| DUBO1355 | | endocast |
| DUBO1356 | | endocast |
| **DUBO1357** | | skull cap *Pithecanthropus,* EDp-1924f (see page 83) |
| **DUBO1358** | | skull cap *Pithecanthropus,* EDp-1924f (see page 81) |
| **DUBO1359** | | skull cap *Pithecanthropus,* EDp-1924f (see page 80) |
| **DUBO1360** | | skull cap *Pithecanthropus,* EDp-1924f (see page 83) |
| **DUBO1361** | | skull cap *Pithecanthropus,* EDp-1924f (see page 82) |
| **DUBO1362** | | skull cap *Pithecanthropus,* EDp-1924f (see page 80) |
| DUBO1363 | | skull tapir 285 |
| DUBO1364 | ⚙ | skull *Oreodon* |
| DUBO1365 | ⚙ | endocasts *Uintatherium* |
| DUBO1366 | ⚙ | brains |
| DUBO1367 | ⚙ | brains |
| DUBO1368 | ⚙ | brains |
| DUBO1369 | | skull cap *Pithecanthropus* |
| DUBO1370 | ⚙ | skull *Protoceratops andrewsi* |
| DUBO1371 | ⚙ | skullsize |
| DUBO1372 | | femur |
| DUBO1373 | ⚙ | skullsize |
| DUBO1374 | | Blanco |
| DUBO1375 | ⚙ | map Telaga Bodas |
| DUBO1376 | ⚙ | map Telaga Bodas |
| DUBO1377 | | Solo man skull in ventral view |
| DUBO1378 | | Solo man |
| DUBO1379 | | Solo man, 1936, no. 10, AnnatI. Utrecht |
| DUBO1380 | | Solo man, 1936, no. 10, AnnatI. Utrecht |
| DUBO1381 | | Solo man, 1936, no. 10, AnnatI. Utrecht |
| DUBO1382 | | Solo man, 1936, no. 10, AnnatI. Utrecht |
| DUBO1383 | | Solo man, 1936, no. 10, AnnatI. Utrecht |
| DUBO1384 | | *Homo modjokertensis* VON KOENIGSWALD 1936 |
| DUBO1385 | | Solo man, 1936, no. 11, AnnatI. Utrecht |
| DUBO1386 | | skull cap *Pithecanthropus,* EDp-1924f |
| DUBO1387 | ⚙ | skullsize Rhodesia man vs australian |
| DUBO1388 | ⚙ | figure canidaea, EDp-1923a |
| DUBO1389 | ⚙ | figure canidaea, EDp-1923a |

|  | |  |
|---|---|---|
| DUBO1390 | | figure canidaea, EDp-1923a |
| DUBO1391 | | endocast |
| DUBO1392 | | geological map of Belgium |
| DUBO1393 | | skull man |
| **DUBO1394** | | antlers *Eucladoceros* Tegelen (see page 112) |
| **DUBO1395** | | antlers *Cervus rhenanus* Tegelen (see page 112) |
| DUBO1396 | | skullfragment with antler *Eucladoceros* Tegelen |
| DUBO1397 | | antler *Eucladoceros* Tegelen |
| DUBO1398 | | antler *Eucladoceros* Tegelen |
| **DUBO1399** | | Trinil 1900 (see page 76) |
| **DUBO1400** | | Trinil 1900 (see page 76) |
| DUBO1401 | | Trinil 1900 |
| **DUBO1402** | | Trinil 1900 (see page 75) |
| DUBO1403 | | skull cap *Pithecanthropus*, EDp-1924f |
| DUBO1404 | | skull cap *Pithecanthropus*, EDp-1924f |
| DUBO1405 | | skull cap *Pithecanthropus*, EDp-1924f |
| DUBO1406 | | skull cap *Pithecanthropus*, EDp-1924f |
| DUBO1407 | | skull cap *Pithecanthropus*, EDp-1924f |
| DUBO1408 | | skull cap *Pithecanthropus* (see page 83) |
| DUBO1409 | | skull cap *Pithecanthropus* |
| DUBO1410 | | skull cap *Pithecanthropus*, EDp-1924f |
| DUBO1411 | | endocast, EDp-1924f |
| DUBO1412 | | endocast, EDp-1924f |
| DUBO1413 | | endocast, EDp-1924f |
| DUBO1414 | | endocast, EDp-1924f |
| DUBO1415 | | skull van Spy |
| DUBO1416 | | skull cap *Pithecanthropus* spec. Dubois |
| DUBO1417 | | skull cap *Pithecanthropus* spec. Dubois |
| DUBO1418 | | Endocast |
| DUBO1419 | X | femur *Pithecanthropus*, EDp-1926m |
| DUBO1420 | X | femur *Pithecanthropus*, EDp-1926m |
| DUBO1421 | X | femur *Pithecanthropus*, EDp-1926m |
| DUBO1422 | X | femur *Pithecanthropus*, EDp-1926m |
| **DUBO1423** | | skull Spy (see page 86) |
| **DUBO1424** | | skull Spy (see page 85) |
| **DUBO1425** | | skull cap *Pithecanthropus* spec. Dubois (see page 84) |
| **DUBO1426** | | skull Spy (see page 5) |
| **DUBO1427** | | skull Spy (see page 85) |
| **DUBO1428** | | skull cap *Pithecanthropus* (see page 84) (broken negative) |
| DUBO1429 | | skull cap *Pithecanthropus* |
| DUBO1430 | | skull cap *Pithecanthropus*, EDp-1924f |
| DUBO1431 | | skull sangiran 1 (VON KOENIGSWALD) |
| DUBO1432 | | plaatje skull Sangiran + stukjes |
| DUBO1433 | | plaatje skull Sangiran + stukjes |
| DUBO1434 | | *Ichthyosaurus* |
| DUBO1435 | | femur *Stegodon*, EDp-1927b |
| DUBO1436 | | cast of Heberlein "skull" |
| **DUBO1437** | | the so-called *Pithecanthropus erectus* (Heberlein "skull") (see page 108) |
| DUBO1438 | | brains |
| DUBO1439 | X | femur *Pithecanthropus*, EDp-1926m |
| DUBO1440 | | femur *Pithecanthropus* |
| DUBO1441 | | femur *Pithecanthropus* |
| DUBO1442 | | femur *Pithecanthropus* afgietsel |
| DUBO1443 | | femur *Pithecanthropus* afgietsel |
| DUBO1444 | | endocast |
| DUBO1445 | | skull Ngandong |
| DUBO1446 | | endocast, EDp-1924f |
| DUBO1447 | | endocast |
| DUBO1448 | | endocast |
| DUBO1449 | | endocast |
| DUBO1450 | | skull Ngandong (broken negative) |
| DUBO1451 | | skull Ngandong |
| DUBO1452 | | skull Ngandong |
| DUBO1453 | | skull Ngandong |
| DUBO1454 | | skull Ngandong |
| DUBO1455 | | skull Ngandong |
| DUBO1456 | | skull Spy |
| DUBO1457 | | skull Ngandong no 11 |
| DUBO1458 | | skull Ngandong |
| **DUBO1459** | | microcephalic skull spec. Dubois (see page 29) |
| DUBO1460 | | microcephalic skull spec. Dubois |
| DUBO1461 | | microcephalic skull spec. Dubois |
| DUBO1462 | | *Homo modjokertensis* VON KOENIGSWALD 1936 |
| DUBO1463 | | *Homo modjokertensis* VON KOENIGSWALD 1936 |
| DUBO1464 | | *Homo modjokertensis* VON KOENIGSWALD 1936 |
| DUBO1465 | | *Homo modjokertensis* VON KOENIGSWALD 1936 |
| DUBO1466 | X | femur *Pithecanthropus*, EDp- |

| | | |
|---|---|---|
| | | 1926m |
| DUBO1467 | | Blanco |
| DUBO1468 | | Blanco |
| DUBO1469 | X | *Pithecanthropus* |
| DUBO1470 | X | *Pithecanthropus,* EDp-1926m |
| DUBO1471 | | skull *Hippopotamus* |
| DUBO1472 | | skull *Hippopotamus* |
| DUBO1473 | ⚙ | endocasts *Uintatherium* |
| DUBO1474 | | skull *Hippopotamus* |
| DUBO1475 | | skull cap *Pithecanthropus,* EDp-1924f |
| DUBO1476 | X | femur |
| DUBO1477 | X | femur |
| DUBO1478 | X | femur |
| DUBO1479 | X | femora |
| DUBO1480 | X | femur |
| DUBO1481 | X | femur |
| DUBO1482 | X | femur, EDp-1926m |
| DUBO1483 | X | femur, EDp-1926m |
| DUBO1484 | X | femur, EDp-1926m |
| DUBO1485 | X | femur van *Pithecanthropus* |
| DUBO1486 | | femur *Pithecanthropus*, EDp-1934a |
| DUBO1487 | | femur *Pithecanthropus*, EDp-1934a |
| DUBO1488 | X | femur *Pithecanthropus*, EDp-1926m |
| DUBO1489 | X | femora |
| DUBO1490 | X | femur |
| DUBO1491 | X | femur *Pithecanthropus*, EDp-1926m |
| DUBO1492 | X | Xray femora |
| DUBO1493 | | Trinil 1900 almost identical to Plate 10, DUBO1399 |
| **DUBO1494** | | Trinil 1900 Plate 9 (see page 75) |
| DUBO1495 | | skull cap *Pithecanthropus* |
| DUBO1496 | | endocast (broken negative) |
| DUBO1497 | | skull cap *Pithecanthropus,* EDp-1924f (broken negative) |
| DUBO1498 | | skull cap *Pithecanthropus,* EDp-1924f |
| DUBO1499 | | skull cap *Pithecanthropus,* EDp-1924f |
| DUBO1500 | | skull cap *Pithecanthropus,* EDp-1924f |
| DUBO1501 | | skull cap *Pithecanthropus,* EDp-1924f |
| DUBO1502 | | endocast, EDp-1924f |
| DUBO1503 | | endocast |
| DUBO1504 | | surface femur *Pithecanthropus* |
| DUBO1505 | | surface femur *Pithecanthropus* |
| DUBO1506 | | surface femur *Pithecanthropus* |
| DUBO1507 | | surface femur *Pithecanthropus* |
| DUBO1508 | | surface femur *Pithecanthropus* |
| DUBO1509 | | femur *Pithecanthropus,* cast |
| DUBO1510 | | femur *Pithecanthropus* cast |
| DUBO1511 | X | femora |
| DUBO1512 | | endocast, EDp-1924f |
| DUBO1513 | | endocast, EDp-1924f |
| DUBO1514 | | endocast, EDp-1924f |
| **DUBO1515** | X | skull röntgen Anna Lojenga (see page 43) |
| DUBO1516 | | femur *Pithecanthropus* |
| DUBO1517 | | femur *Pithecanthropus* |
| DUBO1518 | | recente skull vs microcephalic skull |
| DUBO1519 | | femur *Pithecanthropus* |
| DUBO1520 | | femur *Pithecanthropus* |
| DUBO1521 | | femur *Pithecanthropus* |
| DUBO1522 | | femur *Pithecanthropus* |
| DUBO1523 | | femur *Pithecanthropus* |
| DUBO1524 | | femur *Pithecanthropus* |
| DUBO1525 | | femur *Pithecanthropus* |
| DUBO1526 | | femur *Pithecanthropus* |
| DUBO1527 | | femur *Pithecanthropus,* EDp-1926m |
| DUBO1528 | | femur *Pithecanthropus,* EDp-1926m |
| DUBO1529 | | femur *Pithecanthropus,* EDp-1926m |
| DUBO1530 | | femur *Pithecanthropus,* EDp-1926m |
| DUBO1531 | | femur *Pithecanthropus,* EDp-1926m |
| DUBO1532 | | femur *Pithecanthropus,* EDp-1926m |
| **DUBO1533** | | femur *Pithecanthropus,* EDp-1926m (see page 107) |
| DUBO1534 | | femur *Pithecanthropus,* EDp-1926m |
| DUBO1535 | | femur *Pithecanthropus,* EDp-1926m |
| DUBO1536 | | femur *Pithecanthropus,* EDp-1926m |
| DUBO1537 | | femur *Pithecanthropus,* EDp-1926m |
| DUBO1538 | | premolar *Stegodon* |
| DUBO1539 | | premolar *Stegodon* |
| DUBO1540 | | premolar *Stegodon* |
| DUBO1541 | | premolar *Stegodon* |
| DUBO1542 | | premolar *Stegodon* |
| DUBO1543 | | premolar *Stegodon* |
| DUBO1544 | | premolar *Stegodon* 3381 |
| DUBO1545 | | premolar *Stegodon* |
| DUBO1546 | | premolar *Stegodon* |
| DUBO1547 | | premolar *Stegodon* 3062 |

| | |
|---|---|
| DUBO1548 | premolar *Stegodon* |
| DUBO1549 | *Criptomastodon* |
| DUBO1550 | *Criptomastodon* |
| DUBO1561 | 1.Skull of negro after P. Camper |
| DUBO1562 | 2.Zeus of Otricoli |
| DUBO1563 | 3.Zeus of Otricoli on coin |
| DUBO1564 | 4.Zeus of Otricoli buste in Vatican |
| DUBO1565 | 5.Venus of Arles buste |
| DUBO1566 | 6.Zeus of Otricoli |
| DUBO1567 | 7.crania adult vs child |
| DUBO1568 | 8.crania adult vs child |
| DUBO1569 | 9.dimensions adult vs child |
| DUBO1570 | 10.child with goose |
| DUBO1571 | 11.fresco with wrong dimensions |
| DUBO1572 | 12.human nerve system |
| DUBO1573 | 13.nerve cells |
| DUBO1574 | 14.Purkinje, cortex, cerebrum |
| DUBO1575 | 15.brain ape vs human |
| DUBO1576 | 16.brains several species (Osborne) |
| DUBO1577 | 17.brain centra |
| DUBO1578 | 18.brains human vs 3 animals |
| DUBO1579 | 19.fish brain *Gadus aeglefinus* |
| DUBO1580 | 20.'old' and 'recent' mammals |
| DUBO1581 | 22.brain of Helmholtz vs papua |
| DUBO1582 | 23.Kant |
| DUBO1583 | 24.brain, horizontal cut |
| DUBO1584 | 25.brain parts |
| DUBO1585 | 26.brains comparative |
| DUBO1587 | 27.cerebral cortex comparative |
| DUBO1588 | 28. brains of small dog |
| DUBO1589 | 29.brains of large dog |
| DUBO1590 | 30. brains and eyes comparative |
| DUBO1591 | 31.femurs of *Semnopithecus nasica & S. maurus* |
| DUBO1592 | 32.femurs of *Semnopithecus nasica & S. maurus* |
| DUBO1593 | 33.femurs of *Hylobates syndactylus & H. leuciscus* |
| DUBO1594 | 34. *Lacerta viridus & Varanus* |
| DUBO1595 | 35. long eared bat *Plecolus auritus* |
| DUBO1596 | 36.eel *Anguilla anguilla* |
| DUBO1597 | 37. vipers |
| DUBO1598 | 38.snake (*Naja*) & deer |
| DUBO1599 | 39. *Anguis fragilis* |
| DUBO1600 | 40. *Tapiris indicus* |
| DUBO1601 | 41. *Elephas africanus* |
| DUBO1602 | 42.*Terabelodon augustidens* |
| DUBO1603 | 43.*Meritherium* |
| DUBO1604 | 44.*Elephas indicus* drinking |
| DUBO1605 | 45.*Elephas indicus* bathing |
| DUBO1606 | 46.*Elephas indicus* working |
| DUBO1607 | 47.Ganesha |
| DUBO1608 | 48.*Elephas africanus* cut through skull |
| DUBO1609 | 49.*Elephas indicus* brains |
| DUBO1610 | 50.*Hippopotamus amphibius* brains |
| DUBO1611 | 51.Red faced spider monkey |
| DUBO1613 | 53. *Ateles* var. |
| DUBO1614 | 54. *Ladotrix* Humboldt's woolly monkey |
| DUBO1615 | 55.*Ateles paniscus* |
| DUBO1616 | 56. *Ladotrix* see 54. |
| DUBO1617 | 57.*Microglossus aterrimus* |
| DUBO1618 | 58.*Molge cristata* |
| DUBO1619 | 59. *Hemidactylus turcicus* |
| DUBO1620 | 60.*Rana esculenta* |
| DUBO1621 | 61.*Hyla arborea* |
| DUBO1622 | 62.rat,rabbit and squirrel |
| DUBO1623 | 63. *Tupaia tana* |
| DUBO1624 | 64.seals |
| DUBO1625 | 65.dolphins |
| DUBO1626 | 66.*Balaenoptera rostrata* whale |
| DUBO1627 | 67.*Halicore dujong* |
| DUBO1628 | 68.*Tinca tinca* |
| DUBO1629 | 69.*Alligator niger* |
| DUBO1630 | 70.brains of e.g. *Hesperornis* |
| DUBO1631 | 71.brains and skulls of dinosaurs |
| DUBO1632 | 72.brains of *Ichthyornis & Sterna* |
| DUBO1633 | 73.brains of dinosaurs |
| DUBO1634 | 74.human brains in situ |
| DUBO1635 | 75.human brains |
| DUBO1636 | 76.human brains dorsal |
| DUBO1637 | 77.human brains left hemisfere |
| DUBO1638 | 78.human brains left hemisfere |
| DUBO1639 | 79.'soulblindness' |
| DUBO1640 | 80.cortex map of rabbit |
| DUBO1641 | 81.cortex map of *Cercoleptes* |
| DUBO1642 | 82.brain of *Hapale jacchus* |
| DUBO1643 | 83.cortex map of *Hapale jacchus* |
| DUBO1644 | 84.brain of *Hylobates syndactylus* |
| DUBO1645 | 85.cortex of *Hylobates syndactylus* |
| DUBO1646 | 86.human hemisferes |
| DUBO1647 | 87.lower monkey hemisferes |
| DUBO1648 | 88.cortex map *Pteropus* |
| DUBO1649 | 89.cortex map human |
| DUBO1650 | 90.comparative giant pyramid cells |
| DUBO1652 | 92.brain schematics *Hylobates* |
| DUBO1653 | 93.brain schematics orang utan |
| DUBO1654 | 94.cortex map orang utan |
| DUBO1655 | 95.cortex map *Hylobates* |

| | |
|---|---|
| DUBO1656 | 96.brain schematics *Cercopithecus* |
| DUBO1657 | 97.David & Venus (statues) |
| DUBO1658 | 98.west african green monkeys |
| DUBO1659 | 99.negro monkey |
| DUBO1660 | 100.female black lemur with young |
| DUBO1662 | 102.innervation of muscletissue |
| DUBO1663 | 103.nerve cell |
| DUBO1664 | 104.nerve cell |
| DUBO1665 | 105.*Anabas scandens* |
| DUBO1666 | 106.*Megalobatrachus maximus* |
| DUBO1667 | 107.*Dipus aegypticus* |
| DUBO1668 | 108.*Eliomys nitella* |
| DUBO1669 | 109.*Castor fiber* |
| DUBO1670 | 110.*Thylacinus cynocephalus* |
| DUBO1671 | 111.*Dasyurus viverrinus* |
| DUBO1672 | 112.*Canis familiaris* vs *C. zerda* |
| DUBO1673 | 113.*Canis familiaris* skull |
| DUBO1674 | 114.*Canis zerda* skull |
| DUBO1675 | 115.*Canis zerda* skull |
| DUBO1676 | 116.*Canis familiaris* skull |
| DUBO1677 | 117. *Dolichotis* |
| DUBO1678 | 117a.*Cavia porcellus* |
| DUBO1679 | 118.*Dolichotis?* |
| DUBO1679[a] | 119.*Cricetus* |
| DUBO1680 | 120.*Cricetus* |
| DUBO1680[a] | 120a.*Sciurus* |
| DUBO1681 | 121.*Hydrochoerus* |
| DUBO1682 | 122.*Cavia porcellus* |
| DUBO1682[a] | 122a. *Hydrochoerus* |
| DUBO1683 | 123.*Sorex araneus, Crocidura russulus* |
| DUBO1684 | 124.*Tupaia tana* |
| DUBO1685 | 125.bat |
| DUBO1686 | 126.flying dog (bat) |
| DUBO1687 | 127.mole |
| DUBO1688 | 128.*Potamogale velox* |
| DUBO1689 | 129.weasel |
| DUBO1690 | 130.marter |
| DUBO1691 | 131.*Chiromys madagascarienis* |
| DUBO1692 | 132. *Chiromys madagascarienis* |
| DUBO1693 | 133.*Galeopithecus volans* |
| DUBO1694 | 134.*Hylobates lar* |
| DUBO1695 | 135.*Chrysotrix sciurea* |
| DUBO1696 | 136.skeleton of dolphin |
| DUBO1697 | 137.skeleton of narwal |
| DUBO1698 | 138.brain of *Martes canadensis* |
| DUBO1699 | 139.brain of *Meles taxus* |
| DUBO1700 | 140.brain of *Martes foina* |
| DUBO1701 | 141.skull of *Tillotherium* |
| DUBO1702 | 142.skull of *Brontotherium* |
| DUBO1703 | 143.*Phenacodus primarvus* COPE |
| DUBO1704 | 144.e.g. *Dinoceras mirabile* |
| DUBO1705 | 145.e.g. *Coryphodon hamalus* |
| DUBO1706 | 146.e.g. *Colonoceras agrestis* |
| DUBO1707 | 147.e.g. *Mastodon, Elotherium* |
| DUBO1708 | 148.e.g.*Elephas*, tapir, rhino |
| DUBO1709 | 149.e.g. *Auchenia vicugna* |
| DUBO1710 | 150.*Coryphodon hamalus* MARSH |
| DUBO1711 | 151.*Dinoceras mirabile* MARSH |
| DUBO1712 | 152.*Lorex minulus* (no skin) |
| DUBO1713 | 153.plucked kolibri |
| DUBO1719 | 159.*Canis familiaris* (dwarf) |
| DUBO1720 | 160.*Arctites binturong* |
| DUBO1721 | 52.typical spider monkeys |
| DUBO1722 | 51.Red faced spider monkey |
| DUBO1723 | 53. *Ateles* var. |
| DUBO1724 | 54. *Ladotrix* Humboldt's woolly monkey |
| DUBO1725 | 55.*Ateles paniscus* |
| DUBO1726 | 56. *Ladotrix* see 54. |
| DUBO1727 | 57.*Microglossus aterrimus* |
| DUBO1728 | 58.*Molge cristata* |
| DUBO1729 | 59. *Hemidactylus turcicus* |
| DUBO1730 | 60.*Rana esculenta* |
| DUBO1731 | 61.*Hyla arborea* |
| DUBO1732 | 62.rat,rabbit and squirrel |
| DUBO1733 | 63. *Tupaia tana* |
| DUBO1734 | 64.seals |
| DUBO1735 | 65.dolphins |
| DUBO1736 | 66.*Balaenoptera rostrata* whale |
| DUBO1737 | 67.*Halicore dujong* |
| DUBO1738 | 68.*Tinca tinca* |
| DUBO1739 | 69.*Alligator niger* |
| DUBO1740 | 70.brains of e.g. *Hesperornis* |
| DUBO1741 | 71.brains and skulls of dinosaurs |
| DUBO1742 | 72.brains of *Ichthyornis* & *Sterna* |
| DUBO1743 | 73.brains of dinosaurs |
| DUBO1744 | 74.human brains in situ |
| DUBO1745 | 75.human brains |
| DUBO1746 | 76.human brains dorsal |
| DUBO1747 | 77.human brains left hemisfere |
| DUBO1748 | 78.human brains left hemisfere |
| DUBO1749 | 79.'soulblindness' |
| DUBO1750 | 80.cortex map of rabbit |
| DUBO1751 | 81.cortex map of *Cercoleptes* |
| DUBO1752 | 82.brain of *Hapale jacchus* |
| DUBO1753 | 83.cortex map of *Hapale jacchus* |
| DUBO1754 | 84.brain of *Hylobates syndactylus* |
| DUBO1755 | 85.cortex of *Hylobates syndactylus* |
| DUBO1756 | 86.human hemisferes |
| DUBO1757 | 87.lower monkey hemisferes |
| DUBO1758 | 88.cortex map *Pteropus* |
| DUBO1759 | 89.cortex map human |

| | | |
|---|---|---|
| DUBO1760 | ⚙ | 90.comparative giant pyramid cells |
| DUBO1761 | ⚙ | 91. Indri |
| DUBO1762 | ⚙ | 92.brain schematics *Hylobates* |
| DUBO1763 | ⚙ | 93.brain schematics orang utan |
| DUBO1764 | ⚙ | 94.cortex map orang utan |
| DUBO1765 | ⚙ | 95.cortex map *Hylobates* |
| DUBO1766 | ⚙ | 96.brain schematics *Cercopithecus* |
| DUBO1767 | ⚙ | 97.David & Venus (statues) |
| DUBO1768 | ⚙ | 98.west african green monkeys |
| DUBO1769 | ⚙ | 99.negro monkey |
| DUBO1770 | ⚙ | 100.female black lemur with young |
| DUBO1771 | | jaw fragment Kedung Brubus |
| DUBO1772 | | skull Spy |
| DUBO1773 | | skull Spy |
| DUBO1773[a] | ⚙ | brain size |
| DUBO1773[b] | ⚙ | schematic brain cranium |
| DUBO1774 | ⚙ | ?? |
| DUBO1775 | ⚙ | *Pithecanthropus* |
| DUBO1776 | ⚙ | see DUBO1773[b] |
| DUBO1777 | ⚙ | *Pithecanthropus* |
| DUBO1778 | ⚙ | ?? |
| DUBO1779 | ⚙ | human skeleton |
| DUBO1780 | ⚙ | endocast |
| DUBO1781 | ⚙ | map Solo river Trinil |
| DUBO1782 | ⚙ | aboriginal climbing tree |
| DUBO1783 | ⚙ | Trinil site (van Es, 1926) |
| DUBO1784 | ⚙ | climbing a tree |
| DUBO1785 | | femur distal |
| DUBO1786 | | femur *Pithecanthropus* |
| DUBO1787 | ⚙ | Trinil site (van Es, 1926) |
| DUBO1788 | ⚙ | aboriginal climbing tree |
| DUBO1789 | | femur distal |
| DUBO1790 | | femur |
| DUBO1791 | | femur *Semniopithecus* |
| DUBO1792 | X | femora *Pith.* vs *Hylobates* |
| DUBO1793 | | femur distal |
| DUBO1794 | | femur |
| DUBO1795 | | femur |
| DUBO1796 | | femur *Semniopithecus* |
| DUBO1797 | | femur distal |
| DUBO1798 | ⚙ | femur *Pithecanthropus* |
| DUBO1799 | | femur distal |
| DUBO1800 | | femur distal |
| DUBO1801 | | femur distal |
| DUBO1802 | | femur distal |
| DUBO1803 | | femur distal |
| DUBO1804 | | femur distal |
| DUBO1805 | | femur distal |
| DUBO1806 | | premolar *Pithecanthropus* |
| DUBO1807 | | premolar *Pithecanthropus* |
| DUBO1808 | | premolar *Pithecanthropus* |
| DUBO1809 | | premolar *Pithecanthropus* |
| DUBO1810 | | premolar *Pithecanthropus* |
| DUBO1811 | | molar *Pithecanthropus* |
| DUBO1812 | | molar *Pithecanthropus* |
| DUBO1813 | | molar *Pithecanthropus* |
| DUBO1814 | | molar *Pithecanthropus* |
| DUBO1815 | | molar *Pithecanthropus* |
| DUBO1816 | | molar *Pithecanthropus* |
| DUBO1817 | | molar *Pithecanthropus* |
| DUBO1818 | | molar *Pithecanthropus* |
| DUBO1819 | | molar *Pithecanthropus* |
| DUBO1820 | ⚙ | feet bones *Manis* |
| DUBO1821 | | skull *Manis* |
| DUBO1822 | | skull *Manis* |
| DUBO1823 | | skull *Manis* |
| DUBO1824 | | hand & feet *Manis javanica* |
| DUBO1825 | ⚙ | bones *Manis palaeojavanica* |
| DUBO1826 | ⚙ | bones *Manis palaeojavanica* |
| DUBO1827 | ⚙ | cross-cut femur |
| DUBO1828 | ⚙ | cross-cut femur |
| DUBO1829 | ⚙ | cross-cut femur |
| DUBO1830 | ⚙ | cross-cut femur |
| DUBO1831 | ⚙ | cross-cut femur |
| DUBO1832 | ⚙ | cross-cut femur |
| DUBO1833 | ⚙ | cross-cut femur |
| DUBO1834 | ⚙ | cross-cut femur |
| DUBO1835 | ⚙ | endocast |
| DUBO1836 | ⚙ | cross-cut femur |
| DUBO1837 | ⚙ | cross-cut femur |
| DUBO1838 | ⚙ | plate 8 Trinil |
| DUBO1839 | ⚙ | *??* |
| DUBO1840 | ⚙ | femur distal |
| DUBO1841 | ⚙ | cross cut cast femur dist. |
| DUBO1842 | ⚙ | cross-cut femur |
| DUBO1843 | ⚙ | plate 8 Trinil |
| DUBO1844 | ⚙ | cross-cut femur |
| DUBO1845 | ⚙ | cross-cut femur |
| DUBO1846 | ⚙ | cross-cut femur |
| DUBO1847 | ⚙ | cross-cut femur |
| DUBO1848 | ⚙ | cross-cut femur |
| DUBO1849 | ⚙ | cross-cut femur |
| DUBO1850 | ⚙ | cross-cut femur |
| DUBO1851 | ⚙ | cross-cut femur |
| DUBO1852 | ⚙ | cross-cut femur |
| DUBO1853 | ⚙ | cross-cut femur |
| DUBO1854 | ⚙ | cross-cut femur |
| DUBO1855 | ⚙ | cross-cut femur |
| DUBO1856 | ⚙ | cross-cut femur |
| DUBO1857 | ⚙ | endocast |
| DUBO1858 | ⚙ | cross-cut femur |
| DUBO1859 | ⚙ | cross-cut femur |
| DUBO1860 | ⚙ | cross-cut femur |

DUBO1861 cross-cut femur
DUBO1862 cross-cut femur
DUBO1863 cross-cut femur
DUBO1864 cross-cut femur
DUBO1865 cross-cut femur
DUBO1866 cross-cut femur
DUBO1867 cross-cut femur
DUBO1868 cross-cut femur
DUBO1869 cross-cut femur
DUBO1870 cross-cut femur
DUBO1871 cross-cut femur
DUBO1872 cross-cut femur
DUBO1873 cross-cut femur
DUBO1875 human jaw musculature
DUBO1876 cross-cut femur
DUBO1877 cross-cut femur
DUBO1878 cross-cut femur
DUBO1879 cross-cut femur
DUBO1880 cross-cut femur
DUBO1881 cross-cut femur
DUBO1882 cross-cut femur
DUBO1883 cross-cut femur
DUBO1884 cross-cut femur
DUBO1885 cross-cut femur
DUBO1886 cross-cut femur
DUBO1887 cross-cut femur
DUBO1888 cross-cut femur
DUBO1889 cross-cut femur
DUBO1890 cross-cut femur
DUBO1891 cross-cut femur
DUBO1892 cross-cut femur
DUBO1893 cross-cut femur
DUBO1894 cross-cut femur
DUBO1895 human femur
DUBO1896 *Pithecanthropus* femur
DUBO1897 femur
DUBO1898 femur
DUBO1899 human femur
DUBO1900 femur
DUBO1901 femur
DUBO1902 femur
DUBO1903 femur
DUBO1904 femur
DUBO1905 femur
DUBO1906 femur
DUBO1907 femur
DUBO1908 femur
DUBO1909 femur
DUBO1910 femur
DUBO1911 femur
DUBO1912 human climbing
DUBO1913 human climbing
DUBO1914 human climbing
DUBO1915 human climbing (ED)
DUBO1916 human climbing=DUBO1912
DUBO1917 human climbing
DUBO1918 dental elements
DUBO1919 gorilla mouth musculature
DUBO1920 human climbing=DUBO1914
DUBO1921 *Pithecanthropus* (pre?)molar
DUBO1922 *Pithecanthropus* molar
DUBO1923 *Pithecanthropus* molar
DUBO1924 *Pithecanthropus* molar
DUBO1925 *Pithecanthropus* molar
DUBO1926 jawfragment Kedung Brubus
DUBO1927 cross-cut femur cast
DUBO1928 cross-cut femur cast
DUBO1929 cross-cut femur
DUBO1930 cross-cut femur
DUBO1931 cross-cut femur
DUBO1932 *Pithecanthropus* molar
DUBO1933 *Pithecanthropus* molar
DUBO1934 cross-cut femur cast
DUBO1935 cross-cut femur cast
DUBO1936 *Pithecanthropus* molar
DUBO1937 jawfragment Kedung Brubus
DUBO1938 cross-cut femur cast?
DUBO1939 *Pithecanthropus* molar
DUBO1940 jawfragment Kedung Brubus
DUBO1941 jawfragment Kedung Brubus
DUBO1942 *Pithecanthropus* molar
DUBO1943 *Pithecanthropus* molar
DUBO1944 *Pithecanthropus* molar
DUBO1945 schematic intercranial features La Chapelle skull
DUBO1946 =DUBO1945
DUBO1947 cross section brain cast
DUBO1948 cross section brain cast
DUBO1949 cross section brain cast
DUBO1950 human climbing=DUBO1913
DUBO1951 human lower jaw
DUBO1952 =DUBO1951
DUBO1953 *Pithecanthropus* cranium
DUBO1954 *Pithecanthropus* cranium
DUBO1955 jaw ventral view
DUBO1956 dental elements=DUBO1918
DUBO3901 (30.) adult allometry
DUBO3902 (30.) child allometry
DUBO3903 (27.) cerebral cortex comparative
DUBO3904 (27.) cerebral cortex comparative
DUBO3905 (22.) brain of Helmholtz
DUBO3906 brain gyri
DUBO3907 brains
DUBO3908 (29.)brains of large dog
DUBO3909 (30.)goose brain and eyes
DUBO3910 (30.)human brain and eyes

| | |
|---|---|
| DUBO3911 | (97.)David (Michelangelo) |
| DUBO3912 | (97.)Venus |
| DUBO3913 | skull carnivore |
| DUBO3914 | skull herbivore |
| DUBO3915 | =DUBO3914 mirrored |
| DUBO3916 | skull carnivore |
| DUBO3917 | endocast La Chapelle |
| DUBO3918 | =DUBO3917 |
| DUBO3919 | endocast |
| DUBO3920 | brains |
| DUBO3921 | endocast La Chapelle |
| DUBO3922 | =DUBO3921 |
| DUBO3923 | brain |
| DUBO3924 | brain |
| DUBO3925 | brain |
| DUBO3926 | brain |
| DUBO3927 | cristal-zirkoon |
| DUBO3928 | cristal-zirkoon |
| DUBO3929 | nerve cell |
| DUBO3930 | ?nerve cell? |
| DUBO3931 | (128.)*Potamogale velox* |
| DUBO3932 | rat |
| DUBO3933 | chimp brain schematic |
| DUBO3934 | chimp brain schematic |
| DUBO3935 | brain *Putorius putorius* |
| DUBO3936 | brain *Putorius putorius* |
| DUBO3937 | brain *Mesohippus* |
| DUBO3938 | brain |
| DUBO3939 | brain |
| DUBO3940 | brain |
| DUBO3941 | brain *Procamelus* |
| DUBO3942 | brain *Lama* |
| DUBO3943 | brain |
| DUBO3944 | brain |
| DUBO3945 | brain |
| DUBO3946 | brain |
| DUBO3947 | fissures female brain |
| DUBO3948 | endocast La Chapelle |
| DUBO3949 | endocast |
| DUBO3950 | endocast |
| DUBO3951 | endocast |
| DUBO3952 | endocast |
| DUBO3953 | brain *Patygonus* |
| DUBO3954 | skull *Palaeosyops* |
| DUBO3955 | brain *Cervus* |
| DUBO3956 | (30.)brain & eyes |
| DUBO3957 | *Hylobates agilis* fissures |
| DUBO3958 | brains |
| DUBO3959 | small vs large dog |
| DUBO3960 | gibbons |
| DUBO3961 | (34.)*Lacerta viridis* |
| DUBO3962 | (34.)*Varanus niloticus* |
| DUBO3963 | (131.)*Chiromys madagascarienis* |
| DUBO3964 | (132.)*Chiromys madagascarienis* |
| DUBO3965 | (34.) *Lacerta viridus & Varanus* |
| DUBO3966 | (62.) rat |
| DUBO3967 | (62.) squirrel |
| DUBO3968 | (62.) rabbit |
| DUBO3969 | (132.)*Chiromys madagascarienis* |
| DUBO3970 | (131.)*Chiromys madagascarienis* |
| DUBO3971 | *Homo erectus*>>>modern human |
| DUBO3972 | femur *Pithecanthropus* (64) |
| DUBO3973 | schematic endocast La Chapelle |
| DUBO3974 | endocast Piltdown man…(70) |
| DUBO3975 | endocast (93) |
| DUBO3976 | *Pithecanthropus* cranium (94) |
| DUBO3977 | endocast (95) |
| DUBO3978 | (Plate 12) monument Trinil (91) |
| DUBO3979 | *Pithecanthropus* cranium (98) |
| DUBO3980 | *Pithecanthropus* cranium (99) |
| DUBO3981 | *Pithecanthropus* cranium (100) |
| DUBO3982 | *Pithecanthropus* cranium (101) |
| DUBO3983 | *Pithecanthropus* cranium (102) |
| DUBO3984 | jawfragment Kedung Brubus (103) |
| **DUBO3985** | jawfragment Kedung Brubus (103) (see page 12) |
| DUBO3986 | jawfragment Kedung Brubus (104) |
| DUBO3987 | jawfragment Kedung Brubus (104) |
| DUBO3988 | jawfragment Kedung Brubus (104) |
| DUBO3989 | jawfragment Kedung Brubus |
| DUBO3990 | jawfragment Kedung Brubus (106) |
| DUBO3991 | jawfragment Kedung Brubus |
| DUBO3992 | jawfragment Kedung Brubus |
| DUBO3993 | *Pithecanthropus* molar (107) |
| DUBO3994 | *Pithecanthropus* molar |
| DUBO3995 | *Pithecanthropus* molar |
| DUBO3996 | *Pithecanthropus* molar (109) |
| DUBO3997 | *Pithecanthropus* premolar (110) |
| DUBO3998 | *Pithecanthropus* molar (111) |
| DUBO3999 | *Pithecanthropus* molar (112) |
| DUBO4000 | *Pithecanthropus* molar (114) |
| DUBO4001 | *Pithecanthropus* molar (113) |
| DUBO4002 | *Pithecanthropus* molar (115) |
| DUBO4003 | *Pithecanthropus* molar (116) |
| DUBO4004 | *Pithecanthropus* molar (116) |
| DUBO4005 | *Pithecanthropus* molar (110) |
| DUBO4006 | *Pithecanthropus* molar (117) |
| DUBO4007 | *Pithecanthropus* molar (118) |
| DUBO4008 | *Pithecanthropus* molar (120) |
| DUBO4009 | *Pithecanthropus* molar (119) |
| DUBO4010 | *Pithecanthropus* molar (120) |

DUBO4011 *Pithecanthropus* molar (121)
DUBO4012 *Pithecanthropus* molar (122)
DUBO4013 *Pithecanthropus* molar (122)
DUBO4014 climbing (123)(=DUBO1915)
DUBO4015 *Pithecanthropus* molar
DUBO4016 climbing ()
DUBO4017 *Manis javanica* (127)
DUBO4018 *Manis javanica* skeleton (128)
DUBO4019 *Manis javanica* foot (130)
DUBO4020 *Manis javanica* hand & foot (131)
DUBO4021 *Manis gigantea* (133)
DUBO4022 *Manis javanica* foot (135)
DUBO4023 *Manis javanica* hand (136)
DUBO4024 *Manis tricuspis* (137)
DUBO4025 *Manis temmincki* SMUTS (138)
DUBO4026 *M. palaeojavanica* bones (143)
DUBO4027 Heberlein "skull" (144)
DUBO4028 cast Heberlein "skull" (148)
DUBO4029 Heberlein "skull" (14..)
DUBO4030 cast Heberlein "skull" (149)
DUBO4031 ??? (151)
DUBO4032 bone tissue (152)
DUBO4033 skulls comparative (155)
DUBO4034 dentition human vs gorilla (156)
DUBO4035 pelvis human vs chimp (157)
DUBO4036 reconstruction *Pithecanthropus* skull (158)
DUBO4037 jaw comparative (159)
DUBO4038 cross-cut jaw comparative(160)
DUBO4039 symphysis comparative (161)
DUBO4040 vertebrae human vs gorilla (163)
DUBO4041 jawfragment Kedung Brubus (164)
DUBO4042 jawfragment Kedung Brubus (164)
DUBO4043 femur *Pithecanthropus* (169)
DUBO4044 femur *Pithecanthropus* (170)
DUBO4045 La Chapelle vs *rhodesiensis*(172)
DUBO4046 *Homo rhodesiensis* (171)
DUBO4047 femora comparative (173)
DUBO4048 Rouxs experiment
DUBO4049 Monument Trinil (1907?)
DUBO4050 skull
DUBO4051 skull
DUBO4052 skull
DUBO4053 skull size comparative
DUBO4054 Neanderthal reconstruction
DUBO4055 skull + jaw
DUBO4056 jaw comparative =DUBO4037
DUBO4057 *H. erectus*>human = DUBO3971
DUBO4058 --
DUBO4059 *Pithecanthropus* molar
DUBO4060 comparative feet
DUBO4061 X femur
DUBO4062 *Pithecanthropus* molar
DUBO4063 X femur
DUBO4064 X femur
DUBO4065 --
DUBO4066 --
DUBO4067 *Pithecanthropus* molar
DUBO4068 X femur
DUBO4069 *Pithecanthropus* molar
DUBO4070 *Pithecanthropus* molar
DUBO4071 --
DUBO4072 X femur
DUBO4073 X femur
DUBO4074 X femur
DUBO4075 --
DUBO4076 *Pithecanthropus* cranium
DUBO4077 reconstruction Sussex man
DUBO4078 map: paleo sites in France
DUBO4079 reconstruction *Pithecanthropus*
DUBO4080 *M. javanica* (127) = DUBO4017
DUBO4081 *M. javanica* skeleton (128) = DUBO4018
DUBO4082 *Manis* hand & foot
DUBO4083 *M. javanica* foot (130) = DUBO4019
DUBO4084 *Manis javanica* hand & foot (131) = DUBO4020
DUBO4085 *Manis aurita* HODGE (from skeleton a) female
DUBO4086 *Manis gigantea*= DUBO4021
DUBO4087 *Manis tetradactyla* L.(134)
DUBO4088 *Manis* foot = DUBO4022
DUBO4089 *Manis hand* = DUBO4023
DUBO4090 *M. tricuspis* (137) = DUBO4024
DUBO4091 *Manis sp*
DUBO4092 *Manis temmincki* SMUTS (138) = DUBO4025
DUBO4093 aboriginals vs Venus
DUBO4094 *M. palaeojavanica* bones (143) = DUBO4026
DUBO4095 *M. palaeojavanica* bones
DUBO4096 *Manis sp* skeleton
DUBO4097 *M. palaeojavanica* bones
DUBO4098 Heberlein "skull" (144) = DUBO4029
DUBO4099 cast Heberlein "skull" (149) = DUBO4030
DUBO4100 cast Heberlein "skull" (148) = DUBO4028
DUBO4101 ???
DUBO4102 bone tissue (152) = DUBO4032
**DUBO4103** Thomas Aldous Huxley (see page

32)

DUBO4104 Boucher de Perthes

DUBO4105 Edouard Lartet

DUBO4106 femora comparative (173) = DUBO4047

DUBO4107 Monument Trinil (1907?) = DUBO4049

DUBO4108 jawfragment Kedung Brubus (164) = DUBO4041

DUBO4109 ??? (151) = DUBO4031

DUBO4110 *M. palaeojavanica* footbones

DUBO4111 *M. palaeojavanica* footbones

DUBO4112 pelvis human vs chimp (157) = DUBO4035

DUBO4113 femur *Pithecanthropus* (169) = DUBO4043

DUBO4114 Rouxs experiment = DUBO4048

DUBO4115 *Hylobates agilis* fissures

DUBO4116 symphysis comparative (161) = DUBO4039

DUBO4117 endocast

DUBO4118 endocast

DUBO4119 cross-cut jaw comparative(160) = DUBO4038

DUBO4120 brain size comparative

DUBO4121 1.Skull of negro after P. Camper = DUBO1561

DUBO4122 2.Zeus of Otricoli = DUBO1562

DUBO4123 3.Zeus of Otricoli on coin = DUBO1563

DUBO4124 4.Zeus of Otricoli buste in Vatican = DUBO1564

DUBO4125 5.Venus of Arles buste = DUBO1565

DUBO4126 6.Zeus of Otricoli = DUBO1566

DUBO4127 7.crania adult vs child = DUBO1567

DUBO4128 8.crania adult vs child = DUBO1568

DUBO4129 9.dimensions adult vs child = DUBO1569

DUBO4130 10.child with Goose = DUBO1570

DUBO4131 11.fresco with wrong dimensions = DUBO1571

DUBO4132 12.human nerve system= DUBO1572

DUBO4133 13.nerve cells = DUBO1573

DUBO4134 14.Purkinje, cortex, cerebrum = DUBO1573

DUBO4135 15.brain ape vs human = DUBO1574

DUBO4136 16.brains several species (Osborne) = DUBO1575

DUBO4137 17.brain centra = DUBO1576

DUBO4138 18.brains human vs 3 animals = DUBO1577

DUBO4139 19.fish brain *Gadus aeglefinus* = DUBO1578

DUBO4140 20.'old' and 'recent' mammals = DUBO1579

DUBO4141 Helmholtz

DUBO4142 22.brain of Helmholtz vs papua = DUBO1581

DUBO4143 23.Kant = DUBO1582

DUBO4144 24.brain, horizontal cut = DUBO1583

DUBO4145 25.brain parts = DUBO1584

DUBO4146 26.brains comparative = DUBO1585

DUBO4147 27.cerebral cortex comparative = DUBO1587

DUBO4148 28. brains of small dog = DUBO1588

DUBO4149 29.brains of large dog = DUBO1589

DUBO4150 30. brains and eyes comparative = DUBO1590

DUBO4151 31.femurs of *Semnopithecus nasica & S. maurus* = DUBO1591

DUBO4152 32.femurs of *Semnopithecus nasica & S. maurus* = DUBO1592

DUBO4153 33.femurs of *Hylobates syndactylus & H. leuciscus* = DUBO1593

DUBO4154 34. *Lacerta viridus & Varanus* = DUBO1594

DUBO4155 35. long eared bat *Plecolus auritus* = DUBO1595

DUBO4156 36.eel *Anguilla anguilla* = DUBO1596

DUBO4157 37. vipers = DUBO1597

DUBO4158 38.snake (*Naja*) & deer = DUBO1598

DUBO4159 39. *Anguis fragilis* = DUBO1599

DUBO4160 40. *Tapiris indicus* = DUBO1600

DUBO4161 41. *Elephas africanus* = DUBO1601

DUBO4162 42.*Terabelodon augustidens* = DUBO1602

DUBO4163 43.*Meritherium* = DUBO1603

DUBO4164 44.*Elephas indicus* drinking = DUBO1604

DUBO4165 45.*Elephas indicus* bathing =

DUBO1605

DUBO4166 46.*Elephas indicus* working = DUBO1606

DUBO4167 47.Ganesha = DUBO1607

DUBO4168 48.*Elephas africanus* cut through skull = DUBO1608

DUBO4169 49.*Elephas indicus* brains = DUBO1609

DUBO4170 50.*Hippopotamus amphibius* brains = DUBO1610

DUBO4171 horned toad

DUBO4172 102. innervation muscle tissue

DUBO4173 103.nerve cell

DUBO4174 104. nerve cell

DUBO4175 105.*Anabas scandens* = DUBO1665

DUBO4176 106.*Megalobatrachus maximus* = DUBO1666

DUBO4177 107.*Dipus aegypticus* = DUBO1667

DUBO4178 108.*Eliomys nitella* = DUBO1668

DUBO4179 109.*Castor fiber* = DUBO1669

DUBO4180 110.*Thylacinus cynocephalus* = DUBO1670

DUBO4181 111.*Dasyurus viverrinus* = DUBO1671

DUBO4182 112.*Canis familiaris* vs *C. zerda* = DUBO1672

DUBO4183 113.*Canis familiaris* skull = DUBO1673

DUBO4184 114.*Canis zerda* skull = DUBO1674

DUBO4185 115.*Canis zerda* skull = DUBO1675

DUBO4186 116.*Canis familiaris* skull = DUBO1676

DUBO4187 117. *Dolichotis* vs *Cavia* = DUBO1677 & [a]

DUBO4188 118.*Dolichotis* vs *Cavia* = DUBO1678 & 1682

DUBO4189 119.*Cricetus* = DUBO1679 & [a]

DUBO4190 120.*Cricetus* = DUBO1680 & [a]

DUBO4191 121.*Hydrochoerus* vs *Cavia* = DUBO1681

DUBO4192 122. *Hydrochoerus* vs *C. porcellus* = DUBO1682 & [a]

DUBO4193 123.*Sorex araneus, Crocidura russulus* = DUBO1683

DUBO4194 124.*Tupaia tana* = DUBO1684

DUBO4195 125.bat = DUBO1685

DUBO4196 126.flying dog (bat) = DUBO1686

DUBO4197 127.mole = DUBO1687

DUBO4198 128.*Potamogale velox* = DUBO1688

DUBO4199 129.weasel = DUBO1689

DUBO4200 130.marter = DUBO1690

DUBO4201 131.*Chiromys madagascarienis* = DUBO1691

DUBO4202 132. *Chiromys madagascarienis* = DUBO1692

DUBO4203 133. *Galeopithecus volans* = DUBO1693

DUBO4204 134. *Hylobates lar* = DUBO1694

DUBO4205 135.*Chrysotrix sciurea* = DUBO1695

DUBO4207 137.skeleton of narwal = DUBO1697

DUBO4208 138.brain of *Martes canadensis* = DUBO1698

DUBO4209 139.brain of *Meles taxus* = DUBO1699

DUBO4210 140.brain of *Martes foina* = DUBO1700

DUBO4211 141.skull of *Tillotherium* = DUBO1701

DUBO4212 142.skull of *Brontotherium* = DUBO1702

DUBO4213 143.*Phenacodus primarvus* Cope = DUBO1703

DUBO4214 144.e.g. *Dinoceras mirabile* = DUBO1704

DUBO4215 145.e.g. *Coryphodon hamalus* = DUBO1705

DUBO4216 146.e.g. *Colonoceras agrestis* = DUBO1706

DUBO4217 147.e.g. *Mastodon, Elotherium* = DUBO1707

DUBO4218 148.e.g.*Elephas*, tapir, rhino = DUBO1708

DUBO4219 149.e.g. *Auchenia vicugna* = DUBO1709

DUBO4220 150.*Coryphodon hamalus* Marsh = DUBO1710

DUBO4221 151.*Dinoceras mirabile* Marsh = DUBO1711

DUBO4222 152.*Lorex minulus* (no skin) = DUBO1712

DUBO4223 153.plucked kolibri = DUBO1713

DUBO4224 skulls (cats?)

DUBO4225 skulls (cats?)

DUBO4226 skulls herbivores

DUBO4227 skulls herbivores

DUBO4228 endocasts

| | | |
|---|---|---|
| DUBO4229 | | skull chimpanzee |
| DUBO4230 | | medial view inside skull |
| DUBO4231 | | footprint of mountain gorilla |
| **DUBO4232** | | footprint orang pendak (see page 109) |
| DUBO4233 | | skull large dog |
| DUBO4234 | | skull small dog |
| DUBO4235 | | ED evolution tree |
| DUBO4236 | | monkeys |
| DUBO4237 | | *Tarsius tarsius* |
| DUBO4238 | | monkeys = DUBO4236 |
| DUBO4239 | | comparative feet |
| DUBO4240 | | apes |
| DUBO4241 | | killed gorilla |
| DUBO4242 | | gorilla |
| DUBO4243 | | *Mandrillus leucopaeus* |
| DUBO4244 | | new world monkeys |
| DUBO4245 | | *Ateles* tail |
| DUBO4246 | | comparative feet = DUBO4239 |
| DUBO4247 | | *Cacajao rubicundus* |
| DUBO4248 | | orang utan |
| DUBO4249 | | new world monkeys |
| DUBO4250 | | apes |
| DUBO4251 | | *Tarsius spectrum* |
| DUBO4252 | | hunting *erectus* reconstruction |
| DUBO4253 | | skull humanoid |
| DUBO4254 | | Piltdown skull |
| DUBO4255 | | skull humanoid |
| DUBO4256 | | *Pithecanthropus* cranium |
| DUBO4257 | | skull humanoid |
| DUBO4258 | | -- |
| DUBO4259 | | *Pithecanthropus* molar |
| DUBO4260 | | *Pithecanthropus* molar |
| DUBO4261 | | *Pithecanthropus* molar |
| DUBO4262 | | microcephalic man |
| DUBO4263 | | vertebrae human vs gorilla (163) =DUBO4040 |
| DUBO4264 | X | femur |
| DUBO4265 | | humanoid jaw |
| DUBO4266 | | *Homo rhodesiensis* |
| DUBO4267 | | *Pithecanthropus* cranium |
| DUBO4268 | | *H. rhodesiensis* vs *H. ....* |
| DUBO4269 | | *Pithecanthropus* cranium |
| DUBO4270 | | *Pithecanthropus* cranium |
| DUBO4271 | | ?cranium? |
| DUBO4272 | | skull *Epileptobos* = **DUBO0970** |
| DUBO4273 | | skulls bovines = DUBO0979 |
| DUBO4274 | | Solo river stratigraphy Trinil |
| DUBO4275 | | *Axis lydekkeri* = **DUBO0977** |
| DUBO4276 | | *Stegodon* teeth =**DUBO0937** |
| DUBO4277 | | lower jaws *Stegodon* =DUBO0944 |
| DUBO4278 | | skulls *Bubalus palaeokerabau* = DUBO0980 |
| DUBO4279 | | afstamming hominidea |
| DUBO4280 | | femur distal |
| DUBO4281 | | skull cap of Joe Sibby |
| DUBO4282 | | jaws *Stegodon* = **DUBO0936** |
| DUBO4283 | | *Hylobates* skull |
| DUBO4284 | | Paul Broca coin |
| DUBO4285 | | Plate 2 Kali Ngeto near Kedung Brubus = **DUBO1269** |
| DUBO4286 | | skull homme de la Chapelle |
| DUBO4287 | | humanoid jaws |
| DUBO4288 | | humanoid jaws |
| DUBO4289 | | jaws de la Chapelle |
| DUBO4290 | | diagram endocast de la Chapelle |
| DUBO4291 | | endocast 60 year old female |
| **DUBO4292** | | Rudolph Virchow (see page 30) |
| DUBO4293 | | skull |
| DUBO4294 | | ?? |
| DUBO4295 | | Gustav Schwalbe |
| DUBO4296 | | diagram endocast Piltdown |
| DUBO4297 | | human dispersal |
| DUBO4298 | | humanoid skull |
| DUBO4299 | | Male from Obercassel |
| DUBO4300 | | Female from Obercassel |
| DUBO4301 | | Gustav Schwalbe |
| **DUBO4302** | | Wadjak skull (see page 10) |
| **DUBO4303** | | Wadjak skull (see page 10) |
| **DUBO4304** | | Wadjak skull (see page 10) |
| DUBO4305 | | Wadjak skull - jaw |
| DUBO4306 | | Wadjak skull - jaw |
| **DUBO4307** | | Wadjak skull - jaw (see page 10) |
| DUBO4308 | | endocast *Pithecanthropus* |
| DUBO4309 | | endocast |
| DUBO4310 | | Hand without thumb of *Ateles* |
| DUBO4311 | | *Tarsius spectrum* = DUBO4251 |
| DUBO4312 | | dentition human vs gorilla (156) = DUBO4034 |
| DUBO4313 | | *Homo rhodesiensis* |
| DUBO4314 | | *Homo rhodesiensis* vs Wajak skull |
| DUBO4315 | | *Homo* ?? |
| DUBO4316 | | endocast *Pithecanthropus* |
| DUBO4317 | | endocast *Pithecanthropus* |
| DUBO4318 | | ape skulls comparative |
| DUBO4319 | | endocast *Pithecanthropus* |
| DUBO4320 | | *Homo rhodesiensis* |

## Appendix III: Selection of photographs

Our storyline has made us choose the pictures we have shown so far. There are however many more which we could not fit in for various reasons. Of some it is yet unclear what is depicted, some are in the archives but only as photographs, not as negatives. We present a few here with minimal descriptions, we hope, maybe with help of our readers to further elaborate and complete this list on the website, http://science.naturalis.nl/dubois, hopefully with your help.

DUBO0124 Broken negative of unrecognized location

DUBO0555 Double exposed negative of some event(s) with many people present

DUBO1067 Together with 1062 and 1063: unknown quarry, possibly South-East Belgium??

DUBO0774 Stratigraphy Foulkes Gravelpit Kempston.

DUBO1085 Stratigraphy near Savelsbeek

DUBO1087 Erratic boulder near Moorbeek

Below some photographs Dubois acquired depicting quarries around Tegelen (DUBO2137-49-55-59-61).

These pictures were made in 1912 and judging by the printed text (example below) on the backside made in a series to be sold to tourists visiting the quarries.

DUBO3275 Photograph of Dubois' class at his school in Roermond. Apparently he erased himself from this picture and scratched out some of the eyes of his classmates. Apparently he did not finish this school but did a state exam instead. We are hoping someone will be able to provide us with an undamaged original someday.

DUBO3318 A number of academic photographs like this one with Dubois on it must exist; just a few are in the Dubois collection.

DUBO2042 Terns in the Norangsvalley. Wall facing Southeast-Northwest. (ED)

Enigmatic photograph from Norway (?) of which we hope to clear the significance with your help.

## Acknowledgements

The pictures in this book, when described with a DUBO prefix are all from the Dubois collection at Naturalis. Unless mentioned otherwise pictures were found on the internet. We are grateful to many people and we will list them at the risk of forgetting someone for which we beg their apology. In fully random order we thank Nico Kool and Steef Stijsiger for assisting us in disclosing the Brill archives in the depot of special collections at the University of Amsterdam, Michiel Thijssen, and his collegues at Brill in particular also Sabine Steenbeek, Paul Storm, Ester en Djavid Hadian for judging X-rays, Bert Theunissen, Pat Shipman, Rob Mols for judging Dubois' fotoexperiments, Sabine Hackethal for finding the picture of Margarethe Lenore Selenka, Nelleke Hooijer, Jelle Reumer for providing the photograph of Lorié, Kees de Jong for assisting us in the Artis archives, Gerda Theuns for commenting on the Siwalik pictures, Jan Wieringa for identifying the tree, Jan and Susan de Bruin, Hans Peeters, Conny Canoy-Dankelmans and the Canoy family, John Jagt and Anne Schulp for critical reviews, Ans Molenkamp for invaluable assistance at Naturalis, Berry van der Hoorn, Hanneke Meijer for not throwing us out of her room, José Joordens, Robert Moolenbeek, Bert Boekschoten, Lars van de Hoek Ostende, Erwin Bastiaans, and our respective families for always having to put up with us taking our work home.

Paul Albers & John de Vos

www.ingramcontent.com/pod-product-compliance
Lightning Source LLC
Chambersburg PA
CBHW071246230926
51849CB00001B/243

* 9 7 8 9 0 0 4 1 8 3 0 0 1 *